INANIMATION

Cary Wolfe, Series Editor

(continued on page 319)

INANIMATION

THEORIES OF INORGANIC LIFE

DAVID WILLS

posthumanities **35**

UNIVERSITY OF MINNESOTA PRESS
Minneapolis · London

The University of Minnesota Press gratefully acknowledges financial support for the publication of this book from the Brown University Faculty Development Fund.

An earlier version of chapter 1 was published in German as "Automatisches Leben, Also Leben," trans. Clemens Krümmel, *Zeitschrift für Medienwissenschaft* 4, no. 1 (2011): 15–30. An earlier version of chapter 2 was published in *Mosaic, a Journal for the Interdisciplinary Study of Literature* 44, no. 4 (2011): 21–41. An earlier version of chapter 3 was published as "The Blushing Machine: Animal Shame and Technological Life," *Parrhesia* 8 (2009), http://www.parrhesiajournal.org/. Portions of chapter 4 were published in French as "La technopoétique de l'autre . . . en pointillé," in *Rêver croire penser: autour d'Hélène Cixous,* ed. Bruno Clément and Marta Segarra, 115–26 (Paris: CampagnePremière, 2010). An earlier version of chapter 6 was published as "Raw War: Technotropological Effects of a Divided Front," *Oxford Literary Review* 31, no. 2 (2009): 133–52, reprinted with permission of Edinburgh University Press. An earlier version of chapter 8 was published in *Journal of French Philosophy* 18, no. 2 (2010): 43–64. An earlier version of chapter 9 appeared in Anne Berger and Marta Segarra, eds., *Demenageries: Thinking (of) Animals after Derrida,* 245–63 (Amsterdam: Rodopi, 2011); reprinted with permission of Koninklijke Brill NV.

Published by the University of Minnesota Press
111 Third Avenue South, Suite 290
Minneapolis, MN 55401-2520
http://www.upress.umn.edu

Library of Congress Cataloging-in-Publication Data

Names: Wills, David.
Title: Inanimation : theories of inorganic life / David Wills.
Description: Minneapolis : University of Minnesota Press, 2016. | Series: Posthumanities ; 35 | Includes bibliographical references and index.
Identifiers: LCCN 2015019115 | ISBN 978-0-8166-9882-0 (hc) | ISBN 978-0-8166-9886-8 (pb)
Subjects: LCSH: Life.
Classification: LCC BD435 .W64 2016 | DDC 113/.8—dc23
LC record available at http://lccn.loc.gov/2015019115

Printed in the United States of America on acid-free paper

The University of Minnesota is an equal-opportunity educator and employer.

21 20 19 18 17 16   10 9 8 7 6 5 4 3 2 1

For Emma and Branka
my private lives

CONTENTS

PREFACE

In/animate: what is the appropriate space of separation between those two words? Is there not an absolute distinction to be maintained between what lives and what does not; does not knowledge in general, and scientific knowledge in particular, starting from the eighteenth century, begin there? Or is there rather persistent contamination between the two; at the same time forms or figures of an enlivening of the inanimate, and encroachments of the automatic or inorganic upon the animate? Do we know what *animate* means, any more than *inanimate,* and, presuming we do, would we know how to articulate those two terms vis-à-vis what we call "life"? Aristotle, with his vegetative, sensitive, and rational souls, might take comfort from the avowed inability of contemporary science to define life as a series of positive terms (e.g., programming, self-organization, autokinesis, autopoiesis) and to exclude all counterexamples. But we would no doubt not agree with him any more that the vegetative state of plants renders them lifeless or insensible. And what of the mineral, or the peculiar relations that develop between the human animal and the myriad technological apparatuses that have been developed, most often, precisely in order to render human life easier to live?

The word I will use here to ask these questions is *inanimation.* It is not of my own invention but came into usage, as did the corresponding verb *to inanimate,* in the early seventeenth century (1631 and 1600, respectively). For no less an authority than John Donne, *inanimate* and *inanimation* were the preferred signifiers precisely for the positive senses of "enliven(ing), animat(ing), quicken(ing), infus(ing) life into."[1] To *in*animate was to *en*animate. The privative equivalents, referring to deprivation of life, came later, beginning in 1647 with the verb, which nevertheless remained rare and would soon become obsolete, and in 1784 with the noun, which has managed a longer life.

Inanimation should also be heard in what follows with its own syntactical extension, with its own caesura, referring to *what is inanimate in animation,* documenting the extent to which *the inanimate animates.* It thus borrows less from the lexicon of the cinematic cartoon than from a loose analogy with what is "in suspended animation," to advance the idea of an inanimate (in the contemporary sense) that comes to inhabit a no-man's-land between what doesn't live and what does—an inanimate that imposes itself upon the animate with some type of structural credential. The inanimations that are examined in the chapters to follow may well appear as ontological oddities or monstrosities, especially as they contaminate a literal understanding of living and life by using that semantic framework to refer to patently—or so it is presumed—inorganic instances. But are there, in fact, many words that can lay a lesser claim to literality than *life*? Is there a concept that is any less simple, any more difficult to distill? For the semantic hierarchy of "living" is overlain with so many competing senses, arranged in no coherent order, and at the core, as at the origin of it, there lies conjecture, hypothesis, and presumption, but no certainty. Furthermore, given that uncertainty, what lives appears to be defined to a great extent in contradistinction from what does not live, the animate understood on the basis of the inanimate, which therefore comes to constitute a solid referential kernel, if not nucleus for it, in a similar way to how the inorganic is presumed to constitute the scientific and historical origin of organic life.

The subtitle of this book could have been—in the manner of those participial or interrogative phrases that are designed to illuminate or prescribe on the way to solving a problem—something like this: "Looking for Life in All the Wrong Places" or "Where to Find Life Once the Meaning of It Has Been Irrevocably Called into Question." My aim is not to adjudicate where science continues to debate, nor to claim some creative reconfiguration of the epistemological and discursive formations called "life" that Michel Foucault first developed in *The Order of Things.*[2] But, comparable to both scientific and epistemological projects, *Inanimation* presumes a right to hypothesis or speculation that is at the outside a form of conceptual license: what if, beyond the historical replacement of natural history by scientific biology and a realigned divide between animate and inanimate, organic and inorganic, there were nevertheless room for thinking life outside of that dualism? And what should we make of "loose" or presumed figurative usages of the verb *to live* in instances—such as what "lives on"—that necessarily reflect a different conceptual register for which the appropriate word is lacking? In certain such cases—but, as I hope the studies that follow will show, not in all of them—we can substitute a word like *survive,* but the etymological root of that word belies any absolute emancipation from the concept of life. And if it is easily shown that what lives, in the cases of inanimation that are analyzed here, is incapable of reproduction, what is perhaps less easily determined are its forms of (re) generation or mutation.

Such a conceptual license might function as a response to Jacques Derrida's call to rethink "the limits between the human and the animal, the human and the natural, the human and the technical."[3] Or it might be understood within the context of Gilles Deleuze's "universal machinism: a plane of consistency" beyond animism on one side and mechanism on the other.[4] But beyond the theoretical or philosophical formulations of one or the other of those approaches to the question, what will characterize *Inanimation* is the description or production, or even transvaluation, of a chiasmus between the voice of reason, of science, of academic doxa, and that of rhetorical effect or force, a crossing that will enable the emergence of three unheard of (but neither unknown nor inaudible) forms of life that will go by the names of "autobiography," "translation," and "resonance." In that way, *Inanimation* pursues an endeavor begun with two previous volumes, *Prosthesis* and *Dorsality.*[5] Whereas those books argued the non-purity of the human, insisting on the imbrication of technology within us as far back as, or in whichever direction, we sought to define ourselves as human, *Inanimation* attempts a somewhat converse trajectory: if human life is originarily technological, then what particular artificial, automatic, inanimate, or inorganic forms of life might be identified "within" it or produced by it? The question will be posed much less to account for new and complicated android inventions, however, than to study "phenomena" or operations with which we have long been wrestling and dealing.

Inanimation argues, from Descartes's masked ego in its first chapter to the birdsong of its last, for inanimate life to be recognized in its minimal as much as in its most evident forms. At the same time, it presumes or expects such life or lives to prove itself or themselves only in the event of a performance, as a self-constituting instance, the very constituting instance that is at the origin of what we call life. It is there, where life is presumed somehow to begin, and by means of such a constituting instance, that the inorganic "suddenly" becomes organic; where there was once nonliving, there is now something living, an animate that has come into being by some form of inanimation. Thus the lives I detect or conjure up will perhaps be prosthetic lives, but according to the sense of prosthesis that I have previously maintained, that of an uncanny supplementation of the natural body *by,* but also *in,* its artificial appendage, the body recognizing its own artificiality even as it enlivens its new inorganic member. The prosthetic or prosthetized body is one whose naturality was never intact, never without preparation, organization, articulation, and appeal for its artificial necessities. And my lives will perhaps be dorsal lives, both falling within the imprecise shadow cast by a distinctly reproducible logic of what lives and technologically contrived by the life-forms called human or animal. But the inanimated lives of autobiography, translation, and resonance also enact the paradox of a technology that is presumed to appear at the behest of the human yet that remains unseen, even unenvisageable, in the space of surprise. It is *in front of* us, yet also *in* us to the extent of being *behind* us, like the vertebral edifice around which turns every ensuing frontal orientation. My arguments presume that before there is living—in the sense of what is behind everything that lives, to be negotiated prior to our knowing it to be living, prior to knowing what living means—there is an encounter with the nonliving, with what we might call, presuming to know what *animate* means, the in- or non-animate.

In every case, the discussions to follow remain attentive to how the lives elaborated therein are inanimated by means of language—how language itself generates and self-generates as a privileged form, perhaps *the* privileged form, of inanimate life. That does not reduce to an attention to rhetoric that can somehow be distinguished from a meaning or meanings engendered as some pure, even programmable form of reproducibility, nor is it a matter simply of recognizing that whatever is advanced here is enunciated in the verbal medium. Rather, it involves taking full account of what Deleuze calls the "curious deterritorialization" that consists in "filling one's mouth with words instead of food and noises" (*Plateaus,* 62), words that are inseparable from conceptualization and thinking in general. It means taking that curiosity into account in all its technological force, its inventive momentum, and its extraordinary vitality. Different from all the other technologies of human provenance—indeed, all the objects constituting our constructed material world—whose status as such is never in doubt, and whose prosthetic supplementarity, and converse technologization of the human, has been my constant argument, the words we speak and write inhabit us. They issue from our profoundest depths and never quite leave us, while at the same time falling trippingly off our tongues with the ease of absolute naturality, seemingly as essential and as alive as our breath. Yet, however much they return to us, they can never be ours, and as a result, they both embody and epitomize the intricacy, paradox, potency, motion, fecundity, and whir of life as inanimation.

INTRODUCTION
Doubled Lives

The logic of what lives has its own distinct history, understood as "a history of heredity" in the subtitle of the book published in 1970 by François Jacob, corecipient five years earlier of the Nobel Prize in Physiology or Medicine. *The Logic of Life* is indeed a history, the history of a changing or developing logic, the logic of a changing or developing history of the concept of life and of the living *(le vivant)*.[1] The history that Jacob constructs, of developments in the understanding of life during the modern period leading to the discovery of DNA, is by definition far from closed. Though he presents it as the progressive illumination that we presume scientific discovery to be, he also acknowledges, concluding with the caveat that "science is enclosed in its own explanatory system," that further knowledge concerning how life operates and is defined will lead to, or accompany, further epistemic shifts, which he likens to the emergence of a "new Russian doll" (324).[2]

By Jacob's account, there was no life conceived as such, or at least no organic life, in the seventeenth-century *epistēmē*. Mechanism was accepted as the operative principle across animate and inanimate realms alike. For Hobbes, as for Descartes, as Jacob describes it, "an animal could be considered either as a machine or as an automaton whose limbs twitch like those of a man endowed with artificial life. That was not a metaphor, a comparison or an analogy. It was an exact identification [*une identité*]" (*Logic,* 33; *Logique,* 42). Such an identification was possible because "the living extended without a break into the inanimate. . . . There was as yet no fundamental division between the living and the non-living" (*Logic,* 33). That mechanist presumption, which, "despite its limitations, represented the only attitude compatible with knowledge" (40), would be reinforced by Harvey's (1628) or Newton's (1687) discoveries, and later by the fibers that Bonnet described in 1769 as "infinitely minute Machines" (76) and, in a sense, by the three-dimensional apparatus or "interior mould" that Bonnet's contemporary Buffon imagined in his attempt "to base the properties of living beings on the prevailing laws of physics" (82). By Jacob's account, even the work of Lamarck, one of the primary inventors of biology and, by extension, of life, retains "an old flavour of mechanism [*un fond de vieux mécanisme*], with the regularity and continuity of perpetual or uniformly accelerated motion" (150; *Logique* 167).

However, by the time that Lamarck introduces his theory of acquired characteristics in the early years of the nineteenth century, the major epistemological transformation that constitutes the history of heredity, "by which the living was separated from the inanimate and which established biology as a science" (*Logic,* 152), will have irrevocably occurred. The basis for such a transformation is explicit as early as 1778, in Lamarck's first published work. He begins the text proper of his study of French flora with the following clarification:

> If one observes the different beings that compose the interior structure of our globe, or occupy its outside, one will first remark a large number of bodies composed of raw, dead material which increases by the juxtaposition of the substances forming it and not because of any internal principle of development. These beings are generally called *inorganic* or *mineral beings.* . . . Other beings are provided with organs appropriate for different functions and have the advantage of [*jouissent de*] a very marked vital principle and the faculty of reproducing their like. They comprise the general denomination of *organic beings.*
>
> Those same beings are then divided into two very distinct branches, one containing those that develop and live, but without being endowed with any sensibility and without having the capacity for movement other than that caused by the very organization of the individual, or the action of exterior bodies. These beings are called *vegetal* or *plants.*
>
> The second branch of organic beings is composed of those that share, in addition to development and life, sentiment and spontaneous movement, and these are the animals, among which Man enjoys, with the help of reason, a preeminence that brings him close to Divinity itself.
>
> These inorganic beings have in common with organic beings this point,

that they have the faculty to increase; but they differ in that for the first, such increase occurs by a simple addition or combination of parts, and for the second, by means of development.[3]

Raw, dead matter is here inorganic, first because it is without organs, and second because it is nonvital; third, it is unable to reproduce. Already, it would seem, the criteria by which life will emerge as a positive principle, and as a concept, have their own teleology, which will culminate in the capacity for reproduction. Though the organs that Lamarck describes as lacking in his definition of the inanimate are not of course limited to the organs of reproduction, the vital force of the living can be understood to supersede the mechanics of bodily function in general, and of reproductive function in particular, that is represented by the organs. That vitalist, antimechanist thinking is reinforced by the idea of a movement available to plants, not only as a Newtonian principle (action of exterior bodies), but as issuing from an internal organization, which becomes transcendent spontaneous movement in the case of the animal. What truly lives, from this point on, what is truly animated as animal life, is the "sovereignly automotive," in Derrida's terms, to which we return in following chapters (noting also its paradoxes): "what moves by itself, spontaneously, sponte sua—this is how the living being in general is defined."[4]

As Jacob recounts it, life emerges at the point and in the terms described by Lamarck, and mechanism falters and ultimately collapses into a new divide between living and nonliving. The logic of Jacob's history of heredity, or at least this strand of it that culminates first in Lamarck and his contemporaries, and whose lineage proceeds through Darwin, is that once scientists accepted the principle of organization, the inanimate world came to be distinguished from the inanimate: on one side, there was inorganic, nonliving, inanimate, inert matter, on the other an organic that breeds, feeds, and reproduces (*Logic*, 87). We should therefore infer that classical mechanism is consequently rendered scientifically invalid. But an intersecting strand of that same logic—rather less explicit in Jacob's history but still able to be traced within it—might propose that for life to emerge with the positive characteristics that we now ascribe to it, and certainly for that life to be concentrated in and as a question of reproduction and heredity, a separation had to be made and maintained between the physical or mechanical operations of the animate, on one side, and those of the inanimate, on the other. For it appears less that a universal mechanism did indeed disappear with the emergence of biology than that mechanistic concepts transformed and adapted themselves to the landscape of living versus nonliving, continuing to operate on both sides of the new divide.

The divorce of biology from chemistry and physics, and of things and beings, of the late eighteenth and early nineteenth centuries was in fact short-lived: "The concepts of thermodynamics completely upset the notion of a rigid separation between beings and things. . . . There were . . . two generalizations that brought biology nearer to physics and chemistry: the same elements compose living beings and inanimate matter; the conservation of energy applies equally to events in the living and in the inanimate worlds" (*Logic,* 194). Thus when, in the second half of the nineteenth century, nucleic acid was isolated on the way to understanding the action of enzymes, it was as if "one by one, most of the obstacles raised between organic and inorganic chemistry crumbled" (227); and, by the end of the historical narrative, once the molecular biological analyses of the twentieth century further imposed "the identity of the laws governing the living and inanimate worlds" (266), an identity that was then reinforced by the concept of the program, it would henceforth be presumed that the system of the animal was interchangeable with that of the machine: "The machine can be described in terms of anatomy and physiology. . . . And vice versa, an animal can be described in terms of a machine. Organs, cells and molecules are thus united by a communication network" (253–54). From that perspective, despite the distance traveled by science, and despite the very different scientific facts involved, Jacob's logic appears hobbled by a bizarre feedback loop, an uncanny resemblance between what he accounts for as the most contemporary concept of life, defined by certainty, and the vision of a Hobbes or a Descartes that, by the same account, excluded life in favor of the machine.

One can, I believe, find a similar uncanniness in the historical opposition Jacob develops between the preclassical presumption concerning spontaneous generation and the automatism of reproduction. Belief in spontaneous generation—"in the eyes of the 16th century . . . at least as natural and understandable as generation by seed, if not more so" (*Logic,* 24)—

will have disappeared definitively in the second half of the nineteenth century as a consequence of the theory of evolution; specifically, once chemists such as Pasteur had revealed the role of microorganisms. Granted, Jacob understands spontaneous generation in opposition to reproduction, as having evolved from the classical concept of engendering in the sense of divine creation: "Living beings did not reproduce; they were engendered. Generation was always the result of a creation which, at some stage or other, required direct intervention by divine forces" (19). In the time of Descartes, God came, to some extent, to be replaced by heat and movement, and by the end of the seventeenth century, it was understood that all species produced their own offspring, but the idea of spontaneous generation was to persist, presumably as a form of creationism. The automatism of reproduction, instead, is a cybernetic function, imposed by a "genetic programme" (8), and Jacob uses the term in particular to refer to the impersonal determinism of natural selection; indeed, it might seem that he uses it to preempt any idea of intelligent design: "There is . . . a sorting operation in nature; but it is *automatic*. . . . Among the candidates for reproduction, no intention directs the choice of the elect, which is made *a posteriori* . . . and results only from the qualities and performances of the individuals. . . . A system of this kind can survive only if the 'feedback' loops *automatically* adjust its functioning" (175–76, emphasis added).

A strong deterministic automatism inheres in that basic principle of reproduction: "a bacterium continually [*sans relâche*, "unrelentingly"] strives to produce two bacteria. This seems to be its one project, its sole ambition. . . . For two billion years or more, bacteria—or something like them—have been reproducing [elsewhere it will have been *avec acharnement*, "with desperate eagerness," "ferociously" (*Logic*, 4; *Logique*, 12)]. Structure, function and chemistry of the bacterial cell, all have been refined for this end: to produce two replicas of itself, as well as possible, as quickly as possible and under the most varied of circumstances" (271; *Logique*, 291). Now, without reducing or occluding the major differences between creationist design and materialist determinism, one can argue that there exists a structural similarity between spontaneity and automatism, more than simply rhetorical correspondences among divine causal omnipotence, inexplicable fiat, and the equivalent of a thermodynamic universal law thanks to which "the very notion of causality lost something of its significance" (*Logic*, 200–201). For if references to automatism as such remain relatively rare in Jacob's text, they are nevertheless implied in every reference to the machine or to the program, and they underwrite the whole logic of reproduction.

That logic is finally somewhat tautological: what lives reproduces, what reproduces lives ("Life is born of life, and of life alone" [*Logic*, 126]). Reproduction is without cause, dictated simply by a "cycle of events" (297), but also by "the recopying, with extreme rigour, of a programme" (292). For it is not only a life-form itself that is reproduced but the program that enables that reproduction. Indeed, one could argue that it is therefore the automatic reproduction of a program that constitutes the life that is then able to reproduce itself. Or at least that in a circular system, where "the genetic message can be translated only by the products of its own translation" (30), that is to say, where the reproductive program can be mobilized in its reproducibility only by the reproductions of its own reproduction, there is considerable conceptual and semantic slippage among the terms involved, and that one finds, by no means least among those lapses in rigor, an unwillingness or failure to examine the very concept of reproduction on which the whole of life hinges.

It will be the premise of this book that the whiff of a residual premodern mechanism, which Jacob identified in Lamarck's late-eighteenth-century contributions to what would become biology, is not the only *inanimating* logic of what we now call life. The inanimate does not simply fall away, vehicled by mechanism, into the category of nonlife but continues to operate as an uncanny force across the divide that supposedly protects and defines life. If life emerged in contrast or opposition to the inert machine, it was to develop as another form of machine called a program. And if the living organism were to be defined as automotive and autogenerative, that was possible only within a more general and less categorially defined structure of automaticity. That is not to disprove the logic that Jacob analyzes, nor even to deconstruct the rhetoric of scientific discourse, even though, as I shortly discuss, such a deconstruction opens the question of life to very different lines of inquiry. It is rather, profiting from new forms of the question—as well as everyday usages of the word—to "detect" life on the basis of criteria other than those presumed by the animate–inanimate distinction.

Two different approaches to the question emerge, and my argument tends, according to context, to favor one or the other. The first is consistent with the emphases of my own previous discussions (in *Prosthesis* and *Dorsality*), insisting on a sustained critique of the logocentrism or *physocentrism* of the natural as prior to, opposed to, and distinguishable from the artificial. Recourse to a natural that is presumed to be originary functions as reflex, indeed, a natural reflex that indiscriminately refers to the inanimate physical universe, on one hand, and the animate, particularly human body, on the other. Even the warm day in (northern hemisphere) January that appears just around the corner will not see the end of that *metaphysocentric* presumption, and its critique thus remains, to my mind, an epistemological imperative.

The second approach rejects the oppositional categories of mechanism and vitalism, and the presumed inertness of matter, by developing a conceptual vocabulary and apparatus that refuse to recognize the hegemony of scientific discourse and by producing relations outside normal epistemological or categorial stratifications. From its perspective, the technological will have no more mechanical specificity than will life have organic specificity; just as reproduction comes to be inanimated by a multiplicity or contagion that implies crossing the divide into the inorganic, so life itself will traverse matter by drawing it, irrespective of whether that matter be flesh or metal, into a single inanimated flow.

The first approach produces an inanimation called *lifedeath*. In the course of a seminar in 1974–75, entitled precisely that ("La vie la mort" [Life death]), Jacques Derrida devoted three sessions partially or fully to Jacob's logic of the living.[5] There are two notable strands to Derrida's argument, both of which can be clearly understood in the context of the work he performed in the years 1966–74: an analysis of the conceptual framework of reproduction that Jacob precisely leaves unexamined, studied in particular through the relation of model and copy that the genetic program is designed to regulate, and an unfolding of the textuality (in Derrida's sense) that is thereby implied, and indeed developed explicitly by Jacob (not, of course, in Derrida's sense) in various explanations of the implementation of the genetic program. But beyond that immediate context, as indicated by the title, the work can be read retroactively, through footnoted reference in *The Post Card* (*La Carte postale*, 1980), to the emphases of Derrida's final work and the 2001–3 seminar on *The Beast and the Sovereign*.[6] As the footnote of *The Post Card* explains, the seminar divides into a first series of sessions (sessions 1–6) that concern biology, genetics, and the epistemology and history of the life sciences; a second (sessions 7–10) that turns around Nietzsche and Heidegger's critique of the former's biologism; and a third (sessions 11–14) relating to Freud. The interest in the life sciences might be seen to develop what Derrida delineates in the 1971 essay "White Mythology" as an "epistemological ambivalence of metaphor, which always provokes, retards, *follows* the movement of the concept, [and] perhaps finds its chosen field in the life sciences."[7] His emphasis there is on Canguilhem, who is sometimes referred to in "Life Death," representing those philosophers or epistemologists of life who "implicitly consider that to tamper with the rigorous frontier between concept and metaphor is to compromise scientific objectivity . . . [and so] one abstains from any question concerning the metaphoricity of metaphor and the conceptuality of the concept . . . without suspecting that, far from suffering from an effacement of that limit, science could on the contrary demand a recasting of that division [*partage*] and of the law of that division" ("Life Death," 3:17). But it is Jacob who is subjected to more intense scrutiny inasmuch as he raises "questions concerning the functioning of a certain number of discourses . . . that, however scientific they be up to a certain point, and to that extent their scientificity is not in doubt, nevertheless rely on an uncritiqued, unquestioned operative support—here the notion of production—on which to base all their scientificity" (5:10).

No concept of *reproduction* can, as Derrida sees it, avoid first defining *production* (the seminar also detours through Marx); the concept of reproduction raises not only the question of how copy relates to model but also that of a production presumed to be distinguishable from generation, or creation, or expression. Indeed, "if all the productions of the living—what one calls the productions of the living, and especially the productions of the living entity called man (culture, institution, *technē*, science, biology, texts in the strict sense)—have somewhere as their condition the production of the living as reproduction of itself, and if, on the other hand, the putative 'models' required to understand or know what lives [*le vivant*] are always themselves products or productions of the living, you see not only how entangled this logic is but also how urgent it is to ask: what about self-production and re-production" (5:3)? Questions regarding production, Derrida suggests, are not simply instances of "hermeneutic alchemy" (5:3) aimed at a scientist writing what is evidently a work of vulgarization, obliged to employ an inadequate vocabulary. Jacob repeats from beginning to end of his book that reproduction is what constitutes life: "reproduction acts as the principal operative factor of the living world" (*Logic*, 8; *Logique*, 17); "[living organisms] exist only to the extent they reproduce" (*Logic*, 292). Yet that *reproduction* is defined as a doubling—in the simplest case as the division of a cell—which is not a *production* and not the doubling of a production in any normal sense, certainly not in the sense of industrial or mass production that is more than once a figurative recourse of *The Logic of Life*:

> If analogy is to be used, the bacterial cell is obviously best described by the model of a miniaturized chemical factory. . . . [But] if the bacterial cell is to be considered a factory, it must be a factory of a special kind. The products of human technology are totally different from the machines that produce them, and therefore totally different from the factory itself. The bacterial cell, on the other hand, makes [*élabore*] its own constituents; the ultimate product is identical with itself. If the factory produces, the cell reproduces. (270–71)

The production of identity by the productive apparatus itself is not a form of production that can be easily understood within the normal sense of the word; it does not carry the idea of active force weighing upon matter (as in the production of movement) or of a processing of matter that gives rise to a material outcome (as in a chemical reaction or the production of an artifact or merchandise) but is instead a more or less passive self-constituted event of duplication. Chemistry may well *perform* the reproduction, but it cannot be said simply to *produce* it. Before one arrives at the question of reproduction, therefore, the conceptual quandary of production requires attention, for it will become even more of a quandary, as I discuss later, in the context of what can only be called lifedeath.

The concept of reproduction, as the logic of life-producing self-reproducibility, is for Derrida "hardly conceivable": "An absolute production of self produces a self that is a (living) self only to the extent that, and inasmuch as that originary living production is produced—produces itself—as reproducibility" ("Life Death," 5:1). One can conceive of reproduction only if one somehow accepts that a single cell not only produces two identical cells but at the same time produces reproducibility, as it were, both as mechanism and as concept. For reproduction is always a self-re-production, the self-reproduction of a single cell that we are to understand as programmed to reproduce itself as double, which effectively means that the single cell contains, programmed within itself, the conceptual possibility, nay, necessity, of doubling. "You must multiply" is the cell's programmed command, uttered to an entity that supposedly knows itself only as single: how could it therefore conceive of what that command means, either for its originary state or for its becoming? Reproduction is in that sense something closer to a miracle—at least a conceptual miracle—than a reproduction able to be articulated in the scientific terms that Jacob accredits; it is "a pure event, the absolute event that one supposes to have been produced [produced itself] once only but whose unique production will have consisted in reproducing itself, in dividing itself in order to reproduce itself, in folding back upon itself in order to manifold [*se multiplier*] and therefore disappear as event in the ordinary sense of that word" (6:19).

Hence Derrida's insistence, along lines that are familiar to us from various texts of the period of "Life Death," of an originary reproducibility that deconstructs the simple opposition between single and double and renders highly problematic the concept of reproduction: "In self-reproducing [*le se-reproduire*], neither the *se-* nor the *re-* comes to affect producing [*un produire*] from outside, nor do they happen to a producing that would precede them, a product that would exist before them. What seems to preexist is already a re-product [*un re-produit,* "a reproduced"] as re-product(ion) of itself, a *se-reproduit.* . . . Self-re-production has therefore to be thought other than as what comes to complicate a simple production after the fact. Producibility is from the outset re-producibility, and re-producibility is self-reproducibility" (5:2).

The concept of something called a living cell that nevertheless comes to be constituted as living—according to Jacob's definition—only once it divides and reproduces is fundamentally problematic for understanding what constitutes the living. It is a problematic that Derrida follows as it is played out throughout Jacob's history of heredity, in particular through the logic of the model as a function of genetic programmation and through the rhetorical models—but no longer merely rhetorical nor merely models—that Jacob relies on, such as the factory mentioned earlier and, in far more extensive elaborations, the text.

The operations of heredity are represented as a coded program, as communication network, as message transfer (*Logic*, 1–2), and explicitly as text ("chemical text" [289], "nucleic acid text" [292, 293], repeatedly "genetic text" [290–93]); they require archives, codes, translation and deciphering. But *text*, in those examples and in general, does not easily reduce, no more than does *message*, to a ready-to-hand analogy for the operations of, say, a dioxyribonucleic acid that is itself chemical, a-textual, or meta-textual. Textual operations are not simply represented as functions that DNA performs. Rather, Jacob effectively tells us, heredity *is* a text; the logic of life is the logic of text. On one hand, the "*model* [image] that best describes our knowledge of heredity is indeed that of a chemical message," or "the transformation of a nucleic-acid sequence into a protein sequence is *like* [*ressemble à*] the translation of a message" (275, emphasis added), or "the genetic message *resembles* [*apparaît comme*] a text without an author" (287, emphasis added)—"*like* [*comme*] a text, a nucleic-acid message can be modified by the change of one sign into another" (289, emphasis added)—but on the other hand, those textual analogies or models are analogies or models for the textual operations of the program that transmits a written message. In other words, the model of a genetic text first slips, thanks to rhetorical looseness, into being a genetic text *tout court*: in the detailed descriptions of the culminating subchapter entitled "Copy and Error" (*Logic*, 286–94; *Logique*, 307–14), the model of message, translation and copy, transcription and typesetting error, spoonerism [*contrepèteries*] and rectification, takes over to become what it is analogizing. But beyond that, as Derrida sees it, there can be no extra- or a-textual materiality or scientificity once there is program and message, transmission and communication, code and translation, copying and error. The "genetic message, the programme of the present day organism" (287) that *appears like a text,* is already a text by virtue of being message and program. Heredity and reproduction in Jacob's terms are textual through and through. It is not, therefore, as if Jacob were writing "a text on something said to be outside the text, a-textual, that would form a referent whose nature would be—in its being and in its structure—that it is foreign to textuality but, quite the contrary, he writes a text on a text, on text, and in order to demonstrate, recall, and write that his object has the structure of a text . . . there is no longer anything in the object of his science or his research, as scientific object, that is meta-textual. . . . His ultimate referent, the living, and the productive-reproductive structure of the living is now analyzed as text. Its constitution is that of a text" ("Life Death," 4:1). Derrida will go on to note the serious consequences for a hard, or life, science that shares its object or ultimate referent with philology, literary criticism, or the science of documents and archives: "if there is somewhere a homogeneity (differentiated, but of the same type) between these productions of the living being called man (texts in the narrow sense, calculators, programmers) and the functioning of genetic reproduction, then the opposition between the natural sciences and other sciences loses its pertinence and rigor, and one wonders whether biology can still claim to construct *its* truth, a truth specific to its field" (4:15). Furthermore, if "the science of the living is not one science among others, but also the science implied by all the sciences that determine their object in fields that implicate the living . . . [then] its textualization, the textualization of its object and of its subject, leaves nothing outside it" (4:1–2). Suddenly the famous and infamously misunderstood dictum of the *Grammatology,* "there is no outside to the text," finds a life-shattering instance and a particularly vivid explicitation.

For indeed, the point here is finally less that the logic (of reproduction of life) of Jacob's argument cannot escape its own bind than that the concept of life itself cannot be defined or function outside of what Derrida calls "text," or elsewhere "*différance*" or "trace." Almost thirty years later—as we shall see in chapter 3—when his attention again turns in a concentrated manner to the concept and forms of life, the question will be contextualized in remarkably similar terms, as a matter of there being "*différance* as soon as there is a living trace, a relation of life/death or presence/absence."[8] The logic of reproductive heredity, of the communication of a message whose content or information is not something other than message, communication, or information, is the logic of an irreducible alterity within the same that Derrida calls text: "this textual self-reference, this self-enclosing of a text that refers only to text is nothing tautological or *autistic*. On the contrary. It is because its alterity is irreducible that there is only text . . . and because the set 'text' cannot close upon itself that there is only text, and that the so-called 'general' text (an obviously dangerous and merely polemical expression) is neither a set nor a totality" ("Life Death," 6:4).[9] What the "Life Death" seminar demonstrates with extraordinary clarity is that living cannot be a matter of reproduction without also being a matter of text, trace, and *différance*: text, trace, and *différance* are the facts of life.

The text that Jacob wants to be the figure or model for reproductive heredity is therefore problematic on more than one count: it is a model for the modeling or copying that is reproduction, and what it models is still more text in the forms of program and message, transmission, transcription and translation; and a textualized form of living called the human will go on to produce texts in the strict sense. Furthermore, the message that the text transmits or translates involves how to reproduce and thereby produce the life that, by Jacob's definition, is reproduction. The text therefore signifies not just how life is produced but also—and here we should perhaps say it *performs* more than it *signifies*—how living is produced in its absolute separation from the nonliving.[10] The irreducible alterity that emerges via Jacob's textual model, even considered simply and on its own terms, is a definitive differentiation between nonreproductive nonlife and reproductive life, and the message of the text that transmits the reproductive capacity of the living is also, logically speaking, called upon to transmit that differentiation. For it is only by means of such a differentiation that living can come to be where there was previously only nonlife: life is defined as distinct from nonlife via the performative constitutive operation that is reproduction, a textual operation. In that sense, nonliving text precedes and constitutes life.

It is to explain that differentiating operation that Jacob distinguishes the factory that produces from the cell that reproduces itself, on one side a nonliving artifactual system producing artifacts, on the other a system of the living that reproduces life. He has recourse to that model even if, in the process, he is led to define the production that is a necessary condition for the concept of reproduction as nonliving production, again calling into question the "purity" of a reproductive living. His model means either that the living that reproduces has as a primary element of its definition a nonliving productive machine, or, amounting to something very similar, that there is no production before reproduction and hence no operative model for either reproduction or living, for as Derrida notes, "saying that the machine or the factory produces, and that producing is inferior to reproducing, implies that in fact originary production is on the side of living re-production and that in fact the factory doesn't even produce" ("Life Death," 6:17).

One may, however, note another twist to Jacob's logic of reproduction. In referring to the post-Enlightenment emergence of life as a concept, he had observed that the Diderot and d'Alembert *Encyclopédie* found itself unable in the second half of the eighteenth century to go beyond the self-evidence of a life that "is the opposite of death" (as quoted in *Logic*, 89). The entry continues by arguing that "the smallest life is one from which nothing can be taken away without death ensuing; one can see that in this delicate state, it is difficult to distinguish the living from the dead."[11] At the other end of his history of heredity, Jacob comments on the emergence of death as a function of sexual reproduction: "With bacteria, unlike organisms which reproduce only sexually, birth is not counterbalanced by death. When bacterial cultures grow, the individual bacteria do not die. They disappear as individuals; where there was only one, suddenly there are two.... If this non-life is to be seen as death, it is a contingent death" (297). Sex and death are "the most important inventions" of evolution; indeed, death is a "necessary condition for the very possibility of evolution.... Not death from without, as the result of some accident; but death imposed from within, as a necessity prescribed from the egg onward by the genetic program itself" (309). Although it perhaps becomes, at a certain point, something of a facile exercise to underline all the logical inconsistencies of Jacob's scientifico-speculative argument, it is important in the context of the analyses that follow to note how the *contingent*, "non"-death disappearance of the bacterium via reproduction reinvents itself as the *real* death that ensues from sexual reproduction. For as we shall see in chapter 2, the invention of death implicates not just the life–death relation but also invention–reproduction and, ultimately, inside–outside as the generalized problematic of living itself: the stakes of defining life as a rigorous principle of reproduction, understood as exact replication, turn on—and necessarily stumble over—a system's capacity or incapacity to preserve its own integral intact status. Life will be life in a pure, uncontaminated sense only if it can preserve such a status for itself, replicate itself without exceeding or overreaching itself. If it is tempted to overreach, to reach outside toward something superfluous, as though sexually tempted toward something on the side—"at first it was a kind of auxiliary of reproduction, a superfluous gadget [*un superflu*], so to speak: *nothing obliges a bacterium to make use of sexuality in order to multiply*" (*Logic*, 309; *Logique*, 330, emphasis added)—it may just find death. The bacterium that "knows" only the nonlife or contingent death of multiplication, sacrifice of the one in favor of two, will at some point come to know real life death, real lifedeath, as a threat residing outside it; it will then "decide" to import that death into itself, to have it imposed from within, incorporating it as part of its genetic program—that is to say, to have it become *a*, or *the*, principle of its continued reproductive capacity. The illogicality of such a narrative leads one to ask, first, whether Jacob can in any meaningful sense call the bacterium's reproductive doubling "nondeath" or contingent death, rather than calling it another form of death, a life in death or death in life or lifedeath. And second, as Derrida notes, does not the supplementarity of death render "untenable all the simple oppositions between inside and outside that underwrite what [Jacob's] book says about sexuality and mortality as accidents coming from outside to be inscribed inside?" ("Life Death," 6:12–13). Indeed, is that supplemental logic not "inscribed in the very definition of every system, and even of every living or non-living system" (6:13)?

As Jacob concedes, folding the observation as if seamlessly into his natural law of the genetic program, it was never a matter of automatically programmed intact self-replication and nothing else: "the execution of the prescriptions [the program] contains is subject to specific influences of environment" (*Logic*, 293). But if the effect of the environment is influential, it is not "didactic" (293; *Logique*, 313), especially not in the way of Lamarck's acquired characteristics, something about which recent science is perhaps less absolutist.[12] Indeed, it was always presumed that "the living being cannot be a closed system" (*Logic*, 253), and from the time of Bernard, there was an understanding of a physiological hierarchy of self-protection against exposure to the milieu in accordance with the complexity of the organism. In the simplest organisms, "all the component parts come into direct contact with the surrounding medium," whereas in higher animals, the essential components are not exposed "to the variations and hazards [*aléas*] of the external medium" (187; *Logique,* 204). When that nineteenth-century view is implicitly but effectively revisited in the final pages of Jacob's book, he acknowledges a "tendency to flexibility in the execution of the genetic programme; it is an 'openness' that allows the organism constantly to extend its relations with its environment." Only "in so simple an organism as a bacterium" would the program be carried out with "great rigidity" (308)—only, that is, until that bacterium is tempted by some superfluity.

Reproduction as history of heredity would therefore appear, in the final analysis, to be subject to its own contradictory tension: the logic of life is both program and flexibility, both closed and open, and, as we have also just seen, both replication and replacement, both repetition and contrived difference; ultimately, it is both life and death. As a result, the meaning of reproduction—as meaning of life—is uncannily close to the original biological usage of the word, which, Jacob notes, "was first used to designate the phenomena of regeneration of amputated limbs in certain animals" (*Logic,* 72), a usage that remained current throughout the eighteenth century. The *Encyclopédie* states that "when one cuts the branches of an oak, or fruit tree, or other similar tree close to the trunk the trunk reproduces an infinity of young shoots. By reproduction is ordinarily meant the restoration of a thing which previously existed and which has been destroyed since" (as quoted in *Logic,* 72). The animal example of reproduction was then the limbs of the crayfish.[13] In that sense, reproduction produced something like a natural prosthesis, or at least involved a regeneration of the same as difference. The new shoot or limb grows within the structure of the ablated original and within the structure of the always possible, thus necessarily possible replacement of that original. In other words, the original shoot or limb itself grows within the structure of its potential replacement, and as a consequence, its status as original can be described as contrived or prosthetic. Consequently, one might argue that it is prosthetic contrivance that is being reproduced coextensively with the naturally reoccurring shoot or limb, and within that perspective, any artificial limb that subsequently comes to replace a natural one occupies a structural space that already existed for it, that was already conceiving and anticipating it. Artificial and natural replacements are thus differential possibilities within a shared structure rather than incompatible oppositional elements. As soon as there is reproduction, within reproduction itself there exists the space and structure of artificial or inanimate replacement.[14] A life that is a reproduction that begins as cellular replication with sublational doubling, before evolving to invent a sexual reproduction that means both finding and accommodating an opposite sexual partner and importing death, is constantly revising itself by forms of articulation with various external others that are far from "organic" to it; its evolutionary and hereditary *en*-animations might therefore be understood as types of inanimation.

The life or living that tries to define and conceptualize itself by means of Jacob's far-from-naive logic, appears not to be able to escape "contamination" by the nonliving; it seems that in being conceived, life finds itself in a structurally differential relation with nonlife. In what follows, the paradigm for what is supposed not to live in that relation will be the inanimate, particularly as it is figured in the technological. But forms of inanimation might also be adduced in lesser living beings, such as nonhuman animals, or in nonsentient nature, such as plants, or indeed in the inorganic mineral world. From that point of view, there would be a second, converse approach to inanimation: a nonspecific vitalization of matter to match the animation of the inorganic elaborated so far.

Indeed, as Jacob makes clear throughout his study, strong currents of vitalism both oppose and complement mechanist conceptions of life. In particular, he emphasizes the importance of vitalism for the very emergence of biology in the late eighteenth and nineteenth centuries and the acceptation of life as transcendent organizing principle without which it would be impossible to "maintain the cohesion of the organism, to ensure order in the living being as opposed to the disorder of inanimate matter" (*Logic,* 90). For Bichat, Cuvier, Goethe, and Liebig alike, life was the productive force of resistance against destruction and death. In Jacob's narrative, the "vital force" of that biological moment was later to be transformed into the energy (and entropy) of the thermodynamic age; and the arrival of cell theory and then molecular biology would have definitively "deprived vitalism of its *raison d'être* [*le vitalisme a perdu toute fonction*]" (299; *Logique,* 320). Biology again began to collaborate closely with chemistry, and physics, and vitalism became, as Canguilhem characterizes it, a pejorative term that could only with difficulty protect the study of life "from the annexationist ambitions of the sciences of matter."[15]

That supposed choice, between an eighteenth- and nineteenth-century vitalism that cannot fully emancipate itself from Aristotelian animism, and hard materialist science, or between life as the organism's more or less transcendent cohering principle and an empiricism that comes progressively to be esssentialized in cellular reproductive functions, is precisely refused by a life conceived as rhizome, becoming or immanence in the work of Deleuze and Guattari.[16] Via Spinoza, Leibniz, and Bergson, Deleuze openly embraces at least one lineage of a vitalist tradition that outlives the limits that Jacob wants to ascribe to it. But Deleuze and Guattari are less interested in challenging the narrative that Jacob develops—and the science upon which that narrative relies—than in elaborating a very different logic whose distinction from that proposed by Jacob emerges as a type of composite. In "The Geology of Morals" chapter of *A Thousand Plateaus,* they cite often, and as it were approvingly, *The Logic of Life* and other work, but at the same time they stage something of a rehabilitation of Geoffroy Saint-Hilaire, whose idea of a "general plan" for all organisms within the animal kingdom, and whose emphasis on form over function, was supposedly disproved by Cuvier. As Jacob recounts, "what underlay the idea of a single plan controlling the composition of all organisms was still, up to and including Geoffroy Saint-Hilaire, the old notion of continuity in the living world . . . the chain of beings. . . . This continuity was broken by Cuvier. . . . A breach was opened, not only between beings and things, but also between groups of beings" (*Logic,* 108–10; *Logique,* 123–25). Deleuze and Guattari admit—by twice stating it—that it is difficult to keep one's bearings among the "endlessly proliferating distinctions" (*Plateaus,* 47) of the debates among Geoffroy, Cuvier, and others, but they want to recognize Geoffroy's "topological" thinking in opposition to Cuvier's reflection within Euclidean space, which they later revisit as an opposition between the science of "relations among organs, and among organs and functions" and an epistemology that "goes beyond organs and functions to abstract elements he [Geoffroy] terms 'anatomical,' even to particles, pure materials that enter into various combinations, forming a given organ and assuming a given function depending on their degree of speed or slowness" (*Plateaus*, 254–55). For they are invested ultimately in the destratification of the ontological organization of life that science has canonized, that of the organism as "organic organization of the organs" (158). As Deleuze and Guattari see it, Geoffroy had a presentiment of the direction in which a nascent biological science was leading as he tried to develop the comparative anatomical morphology of what they call "a single Abstract animal, a single machine embedded in the [organic] stratum" (45). Geoffroy's morphology was advanced in opposition to the model developed by Cuvier, which is based on "the hierarchy of functions" that Jacob describes, "a coordinated system that controls the distribution of organs" (*Logic*, 106).[17]

Deleuze and Guattari's vitalism—if that is what it is to be called—presumes a very different set of epistemological conceptions, which imply in turn a different ontological status for the "beings" that constitute life. On one hand, in Spinozist terms, rather than a distinction between inorganic matter and organic life, there is a "single matter" within which what lives *(le vivant)* "performs a transcoding of milieus that can be considered both to constitute a stratum and to effect reverse causalities and transversals of destratification" (*Plateaus*, 336). But if there are stratifications—as well as disruptions of such stratifications—that produce the consistency called life, the conception of such life is by no means exhausted by the category of the organic: "not all Life is confined to the organic strata; rather, the organism is that which life sets against itself in order to limit itself, and there is a life all the more intense, all the more powerful for being anorganic" (*Plateaus*, 503). When Deleuze and Guattari come to name the destratified space across which such a life occurs, they very often refer to it as a plane *(plan)*—"the plane of consistency or composition . . . necessarily a plane of immanence . . . the plane of Nature" (266). It could not be other than the plane of nature, because it is where what transpires, where what comes to pass comes to pass. Conversely, however, as they immediately go on to write, "nature has nothing to do with it, since this plane makes no distinction between natural and artificial" (266).[18] Similarly, the life of, or in, such a plane cannot be confined to the particular corner or segment of it that we call the organic, for it lives by being what transpires, or rather it lives not in *being* anything but in and by transpiring, by occurring transversally, both organically and nonorganically.

The concept of the plane tends to privilege spatiality, which is of course reinforced by the word *plateau* of Deleuze and Guattari's title. But on occasion, *plan* is also understood in the sense of "plan," as organizing principle (265–66), genetic structure (254–55), or architectural design— "urban" planning (368).[19] Now, to the extent that such an organizing, genetic plan or project concerns a "hidden structure necessary for forms" (265), it would "operate as a transcendent unity" (266) similar to the "continuity . . . hidden in the very depths of the living" from which, according to Jacob, nineteenth-century biology was able to break away (*Logic*, 108). The plan is in that sense something quite different from the plane. On the other hand, the French signifier does not make that distinction; it does not so divide the semantico-conceptual field and neither spatiality (plane) nor temporality (plan) accounts sufficiently for how the *plan* is to be conceived. We should therefore also hear the plane of consistency echoing off Geoffroy's "general plan," as the "fixed plan(e) of life, where everything moves, slows down or accelerates" (*Plateaus*, 255; *Plateaux*, 312). It is a plan that departs from itself (without our being able to say what a nonspatial, nontemporal departure means), embracing all that constitutes it, including its failures (without our being able to say what failure means): "it is necessary to pass through fog, to cross voids, to have lead times and delays, which are themselves part of the plane of immanence. Even the failures [*ratés*] are part of the plan[e]" (*Plateaus*, 255; *Plateaux*, 312).

As we know, by the end of his writing, Deleuze was more inclined to develop a simple apposition—with complex implications—between what he called "a life" and "immanence."[20] Life as immanence would be a complicated network of relations that *inhere* rather than *cohere, involve* rather than *evolve*. If there is a temporality to it, that temporality is not linear but represented instead by variations of flow or speed, by mobilities. If one were able to look, from that perspective on life, for anything comparable to a history of heredity such as Jacob recounts, it would certainly not be found in the concept of an organism reproducing according to a determinism that both requires (if there is to be evolution) and ignores (if the species is to survive) its deviations. It would instead be articulated through the logic of becoming. In the section entitled "Memories of a Naturalist" of the major chapter on becoming in *A Thousand Plateaus*, Deleuze and Guattari observe that natural history cannot conceive of "an evolution in the strict sense [*à proprement parler*]. . . . Natural history can think only in terms of relationships (between A and B), not in terms of production (from A to *x*)" (234). Evolution in the strict sense, or "truly speaking," is for them very different from the idea of filiation or descent to which biology tends to limit hereditary processes, which is why "becoming is not an evolution" but something more accurately called "involution," without the sense of regression usually ascribed to that term (238). And similarly, production as Deleuze and Guattari would have it understood (from A to *x*) creates something unforeseen or unknown, nothing that can be predicted by a simple linear causal relation; what gets produced is above all an enfolded or involuted set of relations, "symbioses that bring into play beings of totally different scales and kingdoms" (238). If one puts that sense of production in the context of, and into contrast with, the unexamined conceptual basis for reproduction that Derrida uncovered in Jacob, the logic of life suddenly appears otherwise: a logic of production as contingency, and reproduction as radical difference or multiplicity. Natural history would then become a history of the contranatural, of species born of unnatural nuptials *(noces contre nature)* that traverse the kingdoms of real Nature. Heredity, by extension, emerges as a history of contagion, of "propagation by epidemic" (*Plateaus,* 238; *Plateaux,* 295):

> That is the only way that Nature operates—against itself. This is a far cry from filiative production or hereditary reproduction, in which the only differences retained are a simple duality between sexes within the same species, and small modifications across generations. For us, on the other hand, there are as many sexes as there are terms in symbiosis, as many differences as elements contributing to a process of contagion. (*Plateaus,* 242)

It is not that Deleuze and Guattari deny the science of reproduction, but they do argue for reconceiving its epistemology. When they write most explicitly, after Jacob, about the linearity of the nucleic sequence constituting the reproducible genetic program, it is to draw very different consequences. For Jacob: "what makes the reproduction of molecules so efficient . . . is the existence of a univocal relation between two systems of order: one, the double nucleic-acid sequence that always remains linear and therefore recopies itself without difficulty; the other, the protein sequence that spontaneously and unambiguously converts itself into a specific three-dimensional structure. . . . In the living world, the order of order is linear" (*Logic*, 286). For Deleuze and Guattari, each of those orders of articulation—linear recopying and three-dimensional conversion—represents a shift to another order that functions as a "deterritorialization," an opening to otherness whose subsequent programmatic reterritorialization does not for all that reduce the fact of that "initial" and endlessly repeated displacement: "The alignment of the code or linearity of the nucleic sequence in fact marks a threshold of deterritorialization of the 'sign' that gives it a new ability to be copied and makes the organism more deterritorialized than a crystal" (*Plateaus*, 59). Hence, in a formulation that recalls Derrida's insistence that it is only thanks to *différance* that there can be reproduction, they add that "only something deterritorialized is capable of reproducing itself" (59–60). Unless there is productive potentiality as delinearization, there is no reproduction. That necessary condition obtains all the more stringently in the case of evolution, which takes place by definition as deterritorialization or aberration, by means of a "genetic drift [*dérive*]," a drift that therefore takes life off course, outside the organizing program (*Plateaus*, 53; *Plateaux*, 69): evolution is a production of reproduction's failures—flops, misfires, or screw-ups—that occur as yet another involution on the plane or in the plan of immanence.[21]

The plane of consistency is also called an "abstract machine," which is in turn closely connected to the machinic assemblages *(agencements machiniques)* that mobilize it (or are mobilized across it). It all adds up to what is in one context called a "universal mechanism" (*Plateaus*, 256) and, in another, "the mechanosphere," defined as "the set of all abstract machines and machinic assemblages outside the strata, on the strata, or between strata" (71). The single word "mechanosphere" will be the last sentence of *A Thousand Plateaus,* as it were connecting or assembling the book itself into the plane of what transpires, or conjuring the mechanospheric force to engage the book's machinery. But *machine* will have of course always been one of Deleuze and Guattari's preferred operative words. Their first machines were desiring machines conceived of in order to rewrite a detachable psychoanalytic organicity, and the preceding reference to a "universal machinism" comes at the end of a similar discussion about who or what possesses a pee-pee-maker *(fait-pipi)*.[22] Boys, girls, and locomotives do; chairs don't, because the "organ is exactly what its elements make it according to their relation of movement or rest, and the way in which this relation combines with or splits off from that of neighboring elements. This is not animism, any more than it is mechanism; rather it is universal machinism" (256). Just as the desiring machine tried to show how desire "works"—somewhat in the sense of going to the factory every day—the machine in general reinscribes the supposed "organic organization" between the human organism and its technology within social and political networks. As Ansell-Pearson succinctly explains, Deleuze and Guattari pose the question of the machine "not in terms of the event of the human as a biological organism but rather as directly conceivable in relation to a full social body . . . (for example, the full body of the steppe, the full body of the Greek city-state, the full body of the industrial factory, and so on)."[23] The primacy of the machine derives less from any inherent technicity than from its being a "social or collective machine" (398). It is not, then, a matter of technology as such, of particular "technical elements," but of their being articulated, engaged, assembled, or *agencé* in a set of relations, within a milieu that is not reducible to any purely material contiguity, composed, not of stratified, but of transversal relations: "the principle behind all technology is to demonstrate that a technical element remains abstract, entirely undetermined, as long as one does not relate it to an *assemblage*" (397–98).

A universal machinism is set in train by a movement that is conceived of as relations of speed and slowness among elements that are themselves unformed—except perhaps as molecules (Spinoza's particles)—until becoming bodies affecting other bodies.[24] In similar terms, that machinism will define, and inscribe a given organ—or by the same token a given tool, weapon, or organism—only within a milieu or environment that involves the affecting of bodies one by the other and that is therefore both social and political. A particular assemblage within a given milieu, for example, Von Uexküll's tick, becomes the basis for both an *ethology* and, after Spinoza, an *ethics* (257). A later example—one among many—will explain the emergence of the Greek phalanx as military formation as a libidinal or "passional mutation [*mutation passionnelle*] that drastically changed the relations between desire and the war machine." Men dismount from their horses to participate in an infantry formation, leading eventually to the citizen soldier, and "the entire Eros of war changes" (399).

Beyond that, abstract machines are so called because they "know nothing of forms and substances," but that is also "the rigorous sense of the concept of machine" (*Plateaus,* 511; *Plateaux,* 637; Les machines abstraites ignorent les formes et les substances. Ce en quoi elles sont abstraites, mais c'est aussi le sens rigoureux du concept de machine). The syntax here is ambiguous, saying either that ignorance of form and substance is the rigorous sense of the machine, or that abstraction is, or both. We can perhaps more easily imagine a machine as an abstracting apparatus in the sense that it performs a metamorphosing process, drawing away (Lat. *ab-trahere*) from matter to create something else; and the machine (Gk. *mēchanē*) as contrivance represents some type of abstracted artifact. But a machine that is conceptually rigorous inasmuch as it ignores form and substance takes us to a different concept. Indeed, we are immediately told that such machines "surpass any kind of mechanics. They are opposed to the abstract in the ordinary sense" (*Plateaus,* 511). For if these machines do obey any kind of mechanics, it would have to be a mechanics or hydraulics of fluidity (as in Virginia Woolf's *The Waves*; *Plateaus,* 252), or as we shall see, a more general mechanics of mobility or *motoricity*.

Strangely—at least at first sight—that abstracting specificity to Deleuze and Guattari's concept of mechanicity, the machine's refusal to recognize form and substance, develops through metallurgy. In their concluding summary regarding the plane of consistency in *A Thousand Plateaus,* the authors emphasize the "powerful nonorganic life that escapes strata, cuts across assemblages, and draws an abstract line without contour, a line of nomad art and itinerant metallurgy" (507). Metallurgy has been a focus in the concluding sections of the chapter on the war machine (404–23), constituting "a flow necessarily confluent with nomadism." In those pages, the authors stage something of a debate over the historical role nomads played in the emergence of arms, such as the steel sabre, and inventions, such as the stirrup, which leads them to reinforce how technological innovations spread thanks to itinerant goldsmiths and metallurgists and to observe the prejudices of historians and archeologists against nomadic technological capabilities. But from there they go on to argue more generally that what is lacking is "a sufficiently elaborated concept of technological lineage" (405) and to call for contextualizing metallurgy within a far more complex set of relations. For example, molten steel produced in a crucible requires an assemblage that involves generating sufficient heat, then successive decarburations, then hardness, sharpness, and a polished finish, then undulations and designs that result from crystallization; another weapon, the iron sword, requires a different assemblage (405). Therefore it is a matter not only of recognizing the role of itinerant metallur*gists* but of understanding a certain itinerancy of nonorganic life for which metallurgy is the emblem, the "nonorganic life that... draws an abstract line [of] itinerant metallurgy" (507) just mentioned. They invoke a vitalist mineral nomadism errant within a technological lineage that no longer has anything particularly linear about it, calling it rather a "machinic phylum," borrowing the natural history term for the taxonomic division superior to a class (also used in linguistics for a large grouping of languages). But the machinic phylum is, quite obviously, not limited to the category of the organic, nor is it machinic in the normal mechanical sense, referring rather, as I suggested earlier, to a flow of "matter-movement, matter-energy, matter-flow, matter in variation" (407). It is, affirmatively, a flow of life in the terms already described, reflecting the "technological vitalism" (407), discernible in Leroi-Gourhan or Simondon, that relies on a noninstrumentalist understanding of technology: a technology that does not begin as simple human production or invention and a technology that retrofits the human as *agencement* within a complex technological flow, that therefore technologizes the human.[25]

The nonanimist, nonmechanist universal machinism (*Plateaus*, 256) describes such a movement of matter in general, a flow of life that is reducible neither to what is in any strict sense perceptible nor to the effects of what we understand as pure material contiguity (there being, finally, only one continuous matter). Its mechanism is not, in any case, that of the relation between a human and a tool, referring instead to a type of shifting kaleidoscopic or fractal transversality that, in the informational context, we might now call a network. But the word *machine* retains an epistemologically subversive force when used across the organic–nonorganic divide, for example, to connect the itinerancy of the artisan with a privileged type of mobility that Deleuze and Guattari identify in the metallic. Another emblem—not theirs—of such itinerancy might be the carbon that migrates from minerality to constitute organic compounds and become the basic element of life. What distinguishes metal for them is a similar mobility across the machinic phylum or flow of matter, so that that flow is defined as "essentially metallic or metallurgical" (410). Metal makes explicit what is much less obvious in other transmutations of matter. In the hylomorphic model, for example, something like clay as prepared raw material, existing on one threshold, is formed into a mold on a different threshold; but the stakes of metallurgical transformation are much higher, as the following extended quotation explains:

> In metallurgy . . . the operations are always astride the thresholds, so that an energetic materiality overspills the prepared matter, and a qualitative deformation or transformation overspills the form. For example, quenching follows forging and takes place after the form has been fixed. Or . . . in the case of molding, the metallurgist in a sense works inside the mold. Or again, steel that is melted and molded later undergoes a series of successive decarburations. Finally, metallurgy has the option of melting down and reusing [its] matter. . . . In short, what metal and metallurgy bring to light is *a life proper to matter, a vital state of matter as such, a material vitalism* that doubtless exists everywhere but is ordinarily hidden or covered, rendered unrecognizable, dissociated by the hylomorphic model. Metallurgy is the consciousness or thought of matter-flow, and metal the correlate of this consciousness. . . . Not everything is metal, but metal is everywhere. Metal is the conductor of all matter. (410–11, emphasis added)

Life flows along metal or metallurgical conduits. It extends even, and especially, to that most resistantly inorganic matter, and what appears most solidly and minerally immobile comes to be endowed with deterritorializing motoricity par excellence. For it is that metallic mobility or motoricity, finally, that the machine represents and sets in train. But that is less because we normally think of machines as more or less metallic, or because they consist of moveable parts. It is rather that once we are able to conceive of the metallurgic flow, we will have understood the force of the panmetallic material vitalism that Deleuze and Guattari seek to promote, we will have given to movement an impetus that is irreducible to the perceptible self-displacement of an organism, and we will have definitively freed life from its limited organization within the organic stratum.

Deleuze and Guattari's abstract machine is no doubt irreducible to the sort of prosthetic articulationality that is a frequent conceptual investment within the discussions constituting the three parts of this book. The strategic priority of *Inanimation* remains less a machinic vitalization of the inorganic than a deconstruction of the integrity of the animal organism by means of its inanimate "dependencies." And the concept of life that the book explores is less that of the panvitalist flow of a single matter than a disjunctive set of relations among forms of *protovivance* or *survivance,* cases of minimal attempts at, or ad hoc and post hoc framings of, life on the outside edges of organicity. But the two approaches just outlined often function as two sides of the same coin: they should in any case both be understood as affirmative, affirming what Derrida describes as "the most intense life possible."[26]

Something close to a vitalist approach comes into sharper focus in part III ("Resonance"), where inanimated life will take the form of sonorous echoes that live as relays across disjunctions such as the beating of the heart (chapter 7), the cinematic distinction between sound and image (chapter 8), and recorded birdsong (chapter 9). In those chapters, I return to certain Deleuzian formulations just discussed, such as the machinic phylum, or a metallurgico-musical nexus, or, as developed in his books on cinema, life in and of the image.[27]

Part I ("Autobiography") deals with three lives left behind in writing, or left over in writing, or left over to writing, more or less unwittingly, more or less *autobiographomatically.* From a Descartes whose sybilline formula—*larvatus prodeo*—inaugurates his, and our, modern thinking (chapter 1), to a Derrida who says—perhaps in no less sybilline a fashion— that he can die (chapter 3), with, in between, a Freud who appears to give us a nascent autobiographical psyche in an amoeba (chapter 2), the following three chapters are case studies of what lives *on after,* in the strictest but most problematically tangible sense: words that have been spoken or written, that were more or less designed to last.

Part II ("Translation") deals with a similar operation of afterlife, but in a difference sense. Despite any looseness in the meaning of the word,

Walter Benjamin was very explicit about how literally he understood the relation of life to translation, providing the formulation that has imposed itself as the emblematic dictum for this book: "The idea of life and afterlife in works of art should be regarded with an entirely unmetaphorical objectivity."[28] Translation, nevertheless, translates itself away from any pure literality, as I examine in three chapters that deal with how the breath gets written in Cixous and Celan (chapter 4), with Benjamin's concept of life as translation (chapter 5), and with how extensions of domestic versus foreign space function as zones of war (chapter 6).

So, nine further theories of inorganic, inanimated life.

I

AUTOBIOGRAPHY

1

AUTOMATIC LIFE, SO LIFE

Descartes

In 1619, Descartes is in his early twenties, beginning, one imagines, to fancy himself, yet neither as a mathematician nor a philosopher, rather as something of a writer. He begins to write some notes in Latin in a notebook that will come to light more than thirty years later, after his death. It is there that he makes his famous contribution to what Jean-Luc Nancy has called the "ornamental history of modern philosophy,"[1] by comparing himself to an actor: "Actors, taught not to let any embarrassment show on their faces, put on a mask. I will do the same. So far, I have been a spectator in this theatre which is the world, but I am now about to mount the stage, and I come forward masked" (Ut comoedi, moniti ne in fronte appareat pudor, personam induunt: sic ego, hoc mundi theatrum conscensurus, in quo hactenus spectator exstiti, larvatus prodeo).[2]

However much uncertainty exists concerning the precise dates and order of the fragments from which the statement is drawn,[3] one is to some extent justified in taking it to be Descartes's first published manifesto—not in the sense of the first piece of his writing that was published (the *Discourse*) but rather the first thing that he wrote, chronologically speaking, that subsequently found its way into print. If we exclude letters, legal documents, baptisms where he was a witness, and so on, we may take this *sic ego* to be the first instance, in the context of what has now been published, of a Descartes writing an "I." We can call it his first autobiographical moment. Paradoxically, however, whereas the formal signatory occasions just mentioned, some of which no doubt predated this *sic ego*, historically took on something of a public life, there is no indication that this private journal entry was intended for publication. The notebook was found among his papers in 1650 and subsequently lost. But Leibniz made a copy for himself, also since lost, which nevertheless led to the journal's publication in 1659.

So here is Descartes writing to himself, composing his very own private "I," but, as it would turn out, writing for the first time the writing "I" of a nascent philosopher: he is instituting and constituting the *ego* that will come, once it begins thinking, to define him. Furthermore, he is enacting that as a spectacular performance, declaring it as if from a stage even as he announces that he has not yet mounted the stage. His constative utterance—"actors … put on a mask [*personam induunt*]"— immediately becomes performative declaration: *Sic ego, so I too, I now pronounce myself an actor, I come forward masked.* It would take us a long time to get beyond the intricacies of that performance, beginning with its complicated proleptic temporality ("I am on the point of becoming an actor and may or may not have already put on my mask"), as well as the transformation itself from spectator to actor, mediated by the complicated metonymies of the mask or *persona,* a word for a mask in the form of both a visor or camouflage and the person of a person playing a role. In a sense, everything that follows here will involve trying to understand this rehearsal of Descartes's, this preview to the performance of his life, to his performance of a lifetime.

How can we understand the status of this first written *ego*? What sort of consistency can we presume to be at work in the *ego larvatus* of these early writings: an *ego* who jots down some thoughts in Latin eighteen years before declaring "*Je pense*" in the *Discourse* (1637), or an *ego* enunciated twenty-two years before becoming the *ego sum* who declares that it doesn't yet know what that *ego* is in the *Meditations* (1641)? I pose that question less in the sense of a philosophical conundrum, such as the precise one that Descartes dedicated his thinking to solving, cogently analyzed by Jean-Luc Nancy in *Ego Sum,* even though what I develop here cannot not, at various points, intersect with that problematic. I pose the question less in that sense than in terms of the functioning of an autobiographical animal that seems each time to be at work, that seems to come automatically into play, a *zoon automaton graphikon that is I,* here expressing itself as such for the first time.

The mask (Lat. *larva*) that Descartes puts on is in the first instance something horrific. The Latin word refers to a frightening, spectral mask, and before that to the specter or ghost itself. Though we know there is nothing innocent about donning a mask, we should keep in mind the full duplicity that it involves. If here the actor is presumed to put on a neutral or neutralizing mask to cover his embarrassment, once there is masking, we can never know with confidence what the relation is between the mask and what it covers: opposition (a nice mask for a frightful face, or vice versa), mimicry (a differently nice mask or a differently frightful mask), or intensification (a nice mask for a nicer face, a frightful mask for a more frightful face; not to mention dis-*simulation* itself—a mask to hide another mask). The mask is always a specter to that extent, to the extent that the relation between mask and face remains haunted by that uncanniness, by what, in any doubling, remains unknown. When Linnaeus coined the word *larva* in 1691 for its current zoological sense, he was thinking of how the larva masks its future metamorphosis, as it were, seeking to rid the word of its uncanniness by inserting the masking in a predictable teleological chain, always pointing forward to the butterfly. But Linnaeus would also have set loose a specter from the future to haunt what Descartes writes circa 1619, positioning this *ego larvatus* as a barely metamorphosed larval subject, something of the "aborted Cogito" to which Deleuze refers, after Ricoeur.[4]

Sic ego, he writes, "so I," "thus it is with me." "Actors, taught not to let any embarrassment show on their faces, put on a mask. I will do the same [*sic ego*]. So far, I have been a spectator on this theatre which is the world, but I am now about to mount the stage, and I come forward masked." The Cambridge English translation makes *sic ego* a sentence of its own, explicitly relating it, in syntactical terms, to putting on a mask. That is how the Latin works, at least if we accept the punctuation that was added at the time of publication (Ut comoedi, moniti ne in fronte appareat pudor, personam induunt: sic ego, hoc mundi theatrum conscensurus, in quo hactenus spectator exstiti, larvatus prodeo), for the words *sic ego* begin the series of clauses about being a spectator, mounting a stage, and coming forward masked. One might more literally have translated it thus: "Thus/ So I, on the point of mounting the stage of the theatre of the world in which I have so far been a spectator, come forward masked." Syntactically, *sic ego* links directly with *larvatus prodeo*: "Thus I . . . advance masked." But, semantically speaking, *sic ego* cannot not also refer back to the actor's embarrassment, for the implication is that "embarrassed actors put on a mask to hide their embarrassment, so it is with me who advances masked." In other words, Descartes's first ego, some time before it will become an *ego cogito* or an *ego sum* or an *ego existo,* is an *ego larvatus* (a masked self), but even before that, it is an *ego pudeo* or *pudendus,* an embarrassed or ashamed self. Before being masked with its own mask, which is not its own mask, it is masked with another mask, which *is* its own mask, more properly proper to it, namely, its shame or embarrassment. Descartes, at something of a beginning of his writing career, inasmuch as he fancies himself as an actor, decides not to wear the mask of an actor without having first consented to wearing the mask of embarrassment.

Now of course we understand embarrassment as something that we precisely don't consent to. We presume the mask of embarrassment to be a naturally occurring mask and the actor's mask to be an artificial, prosthetic one. We can take blushing to be emblematic of the facial con-*figuration* of embarrassment that Descartes is talking about here, although it could be a grimace or some other tic. Blushing, as paradigm for this sort of *discountenancing,* is something we blame on adrenaline, on the rather unsympathetic operations of our sympathetic nervous system. It is a system as essential to our existence as breathing: *I blush therefore I am* in the same way as *I breathe therefore I am,* in the same way as *I panic therefore I am*—long before *I think therefore I am.* Blushing is simply part of life. Indeed, if we had to think about blushing or breathing, we would likely never get around to doing much thinking. We are just like other animals in that regard; our nervous systems have evolved to the point where we breathe without thinking. Except that, paradoxically, other animals breathe without thinking, like us, and panic without thinking, flee without thinking, like us—indeed, their adrenal glands are in general much more active than ours—but apparently they would never think to blush; not only would it never enter their heads but it wouldn't even enter their physiology. That makes blushing, as Darwin noted, "the most peculiar and the most human of all expressions. Monkeys redden from passion, but it would require an overwhelming amount of evidence to make us believe that any animal could blush."[5] That is rather ironic given that blushing returns the human to a type of animal, female, or infantile docility or vulnerability, that of the blushing virgin, for example, or of adolescents in general.[6] For if blushing is a particularly human form of automatism, that is because it is derived from a particularly human form of shame, derived in turn from a particularly human situation, namely, nakedness. At least that is what is presumed by the whole etymologico-mythological network within which Descartes's verb for being embarrassed operates: *pudor* signifies the shame of exposing one's *pudenda* or sexual organs; it refers back to our original fault or fall, into nakedness and into technology.[7] However much shame is related to the pure life of the spontaneous blush, the automatism of blood rushing to the face, it is also a function of the originary technicity that is the origin of technology; we can identify in it one of the "first" moments of the technological drive. Simultaneously, and automatically, as we recognize our nakedness, we cover it for shame—or *with* shame—and so put on our first layer of "clothing." In that originary prosthetic moment, the supposed natural blush opens the structure of the artificial supplement. Technology begins thus, with a red face, with the particularly human "decision" to blush, with our bioengineering of adrenal functions that starts a chain reaction: hands covering the groin, a fig leaf quick, please, or better still, sew me a loincloth, but make it fast— and it only gets more complicated from there on. As Derrida has written, "clothing derives from technics. We would therefore have to think shame and technicity together, as the same 'subject.'"[8]

The shame of nakedness is something we should rightly call a technique, something that, for being neither the first nor last instance of it, takes us out of nature—something that humans alone have added to their physiological repertoire. The technology that produces clothing begins long before the first stitch, in the contrivance by which we require ourselves to conceive of nakedness so as to be able to conceive of non-nakedness, or vice versa, for, as I explain further in chapter 3, the very dialectical relation, and hence thinking itself, has itself to be contrived in the same process. I therefore again read some uncanny prescience in this first relating of the *ego*—as it were before the fact of the thinking *ego* but necessarily inscribed in its evolution or emergence—to shame; something of a conceptual quandary that will persist once Descartes mounts the stage to play his leading role in modern philosophical history. It would require us to accept either that there is indeed already thinking in blushing—the "decision" just referred to—or that an automatism such as that of shame or embarrassment, a natural automatism like breathing, continues to function within our presumed autonomous conscious thinking. Either blushing is something we have dreamed up as a result of thinking shame, bioengineering the adrenal gland with a retrofitted conceptual plug-in, or we possess a type of sympathetic thought system that never gets fully interrupted by the activation of the cogito—not just a sympathetic nervous system and a cogito functioning in parallel but one inhabiting the other.

Such is Descartes's first *ego* inscribed here: *ego sic,* just as it is, as it is with it. Being an actor, it puts on a mask to cover the mask that it has already, supposedly uncontrollably, put on—an artificial mask (persona) to cover a natural one (blushing). Putting on the second, really artificial mask would be the act of a *cogito,* even though that doesn't yet exist, whereas the first, totally natural red-faced mask would represent the automatism of a "physiological" *ego.* But we are talking about someone or something that, before having the mask of embarrassment thrust upon it, and before putting on a mask to cover that embarrassment, already wears the mask of acting, having chosen to adopt the profession of one who borrows another persona. The *ego* of *sic ego* refers back not only to the fact of being embarrassed but also to the decision to be an actor, which is what has induced the embarrassment. "Actors" is in fact, according to the logic we adopted earlier, Descartes's first substantive written word (*Ut comoedi...*), and acting is the first constitutive attribute of his *ego*; the fact of acting, adopting a persona, is his first mask. Before donning a mask, he dons a persona, that of an actor who will subsequently adopt other personas and the masks that go with them. As a result of choosing to be an actor, he writes, which is as it is with me, I find myself in a situation where I feel shame, and so blush, requiring therefore a mask. I act, therefore I feel shame, therefore I blush, therefore I put on a mask.

The actor's embarrassment, his *pudency,* is not something in his head or heart, or wherever such a feeling is situated; it is a shame that acts out on his face, which is what has led me to refer to blushing as a mask. But that begs the question of knowing whether in fact blushing serves to hide shame or to express it. We read blushing as signifying shame, but if the sympathetic nervous system were indeed sympathetic, if it were to function as the survival technique it is supposed to be, it would be helping me to avoid showing the shame that I feel; it would be masking, not expressing, my shame. The automatic reactive linear logic of the system seems therefore to have broken down. In the case of other adrenal functions, the hormone encourages us to preserve ourselves by fighting or fleeing; here we are forced into defenselessness, or at least into a more complicated form of defense. Within the natural mask that is blushing, we would thus have to recognize something like a failed attempt at a different kind of mask, which would make of blushing a doubly masked affair, both purely adrenal (oh, shit! pump blood!) and value laden (my bad, I admit it, see for yourself, I'm not trying to hide it). It is as if we could identify a corollary to the thinking blushing posited earlier, structurally imbedded within this physiological mechanism. In the blush, there would be a failure or breakdown of the natural hormonal process and a contrived attempt to cover it: instead of (or as well as) simply increasing blood flow to facilitate defensive action, the adrenal response is somewhat counterproductive, stimulating a facial cutaneous blood flow that would seem to be least needed and inducing a type of confusion between the purely physiological and the psychological (presuming such a distinction can be maintained). The blush interrupts the reactive nervous process and complicates it with a whole set of culturally determined factors: adolescent sensitivity, like seductive coyness, no doubt related (consciously or unconsciously) to socialization as sexualization; variation in skin pigmentation as a function of racial difference related to color–emotion associations, and so on.[9] Blushing thus reveals itself as an epidermal or membranous site of articulation between the quick of cutaneous matter and the nonphysiological or nonanimate, a natural prosthesis and the sign of an imbrication of animation and inanimation that we might presume to have been always already in play.

Sic ego, thus it is with me: I act, therefore I feel shame, therefore I seek to mask that shame, therefore I fail to do so and have to make do with the red mask that my body hands me, therefore I put on a mask. I put on the mask of acting, the mask of shame, the mask of a blush designed to hide that shame, the mask of a blush that expresses that shame, and finally the mask of an actor. How many masks does that make, finally? In what precise order? Where do they stop being natural and begin to be contrived? And did we even agree about why actors had to worry about showing their embarrassment in the first place, or what exactly they were ashamed of? Was Descartes referring only to his own schoolboy dramatic efforts at the Collège de la Flèche, as Charles Adam seems to think? Or, as Gouhier contends otherwise, was the philosopher alluding to the stage fright that haunts all actors?[10] If the latter, are they afraid of blushing their very own personal offstage nonacting blush at the wrong moment, having lapsed out of their role, and so be guilty of bad acting, if not of acting badly? Or rather, of playing their role so well, being so well into character, that they blush the blush any normal human would blush, whereas the stage directions call for the character to maintain his duplicity or hide his shame? Can we distinguish the personal accidental blush of a normal human functioning as an actor who happens to be playing the part of a nonblushing character from the personal accidental blush of that character as played by a normal human functioning as an actor? Or, more generally, is the very fact of acting, by definition, a shameful business that actors have to hide? Within the Greek and Western tradition, acting emerges from bacchanalian celebrations. A *komoidia* is the song of revelers disguised as lascivious satyrs, the unruly dithyrambic dancers (Gk. *komoi*) who became a chorus, in counterpoint to which there eventually developed an individual role-player. Such a tradition, and link between acting and orgiastic practices, would seem to take us back to the embarrassment of a type of nakedness or sexual exhibitionism as the reason for the shame of acting. Actors, in that respect, would be ordinary humans, their shame ordinary shame, and their masks ordinary clothing, originary technics.

Descartes's first private meditation, therefore, before he develops his *Rules for the Direction of the Mind,* before he writes his *Discourse on Method* or his public *Meditations,* consists in offering this constative utterance: actors wear masks. Which means on one hand "there are actors, therefore there are masks," but also, more basically, "there is acting, there is masking," without any "therefore," no adverbial conjunction, just a pleonasm or tautology without causal consecution: acting (there) is, masking (there) is. Long before any *I* gets to think or to be, in the very beginning, acting is what is, masking is what is. From the opening of such an abyss of prosthetic supplementation, whether pretense (the addition of a disguise) or orthopraxis (the rectification of a failing), there follows the whole apparatus of *personations* that I have just described. But it by no means ends there. Our actor masked with acting, shame, blushing, and a mask is, at the precise point at which he appears in the writing, not yet an actor. He is still a spectator who wishes or plans to be an actor, or he is an apprentice actor, someone preparing to be an actor, as yet only the pale imitation of an actor, a spectator playing the role of an actor, pre-acting acting, borrowing the mask of an actor yet not blushing, it seems, on account of his lack of experience, and not yet blushing, it seems, the blush of a real actor. He waits offstage, out of the limelight, more precisely below the stage, hiding as it were, wearing the mask of darkness, waiting for the cue that will come entirely at his own prompting, for his ego is already announced as being-in-ascendant *(sic ego ... conscensurus)*; he is on the point of mounting the stage and stepping out. Lowly spectator that he currently is, he will shortly borrow the persona of an elevated actor. Consonant with the whole, seemingly endless thematics of masking, there exists in this passage another semantics, of moving or putting oneself forward and upward, of a progression that is a self-promotion and a self-projection or propulsion: I am moving out from the orchestra seats to the point from which I shall shortly climb onstage so as to advance masked. I am at the same time moving upward and forward and holding something in front of me.

Two things should be accentuated here: first that the prosthetization of the *ego* that takes place by means of the complicated effects of masking that I have outlined is doubled by the putting out, up, and forward *(conscensurus/prodeo)* of this self-promotion. To *prosthetize* is also to *propose* (indeed, to *prostitute*), a form of advancing or projecting oneself, by which the self readjusts itself outside of itself, as I have argued before.[11] The thematics of masking and progressing culminate in the *larvatus prodeo*: "I advance masked" means that I advance by advancing my mask and also that I mask myself by advancing in the sense of transforming and transcending my (former) self; I go forward putting a mask in front of me; I put both my mask and my body out in front, my mask out in front of my body and my body out in front of itself. As soon as I move up and out in front of myself, before I put on the transformative mask, I am moving myself into otherness, something that is explicitly reinforced here by the metamorphosis from spectator to actor and from an inferior, or common, situation to an elevated one. Concentrated in the *ego larvatus prodeo* is thus an *I* that reveals itself as originary prosthesis, an *I* that moves forward out of itself and so opens the space and structure of masking, masks itself with its own prospective otherness, on the basis of which it is capable of putting on a mask. But it is also a dorsal *I*, always behind itself not just because its ontology is progressive or successive, substitutive or supersessive—it is ascending, advancing, it *is* by ascending, advancing with respect to itself—but also because, according to the complicated logic we have just seen, the technology of prosthetic supplementation does not begin with the mask that is named as such but is already in operation from the first: where there is acting as masking, there is already covering, and that *behind* any and every manufactured artifactual disguise. And from the first, what is before and behind anything else is acting and masking.

The second point to emphasize is the force of a certain verticality, or verticalization, that functions in association with masking: Descartes's *I* will ascend or mount the stage, achieve a superior uprightness or erection. That is yet more cause for shame and masking, perhaps; indeed for both masking and nonmasking, for we know the complicated play of the codpiece, like the noses of *commedia del arte* masks, as both signifier of and cover for virility, tumescence, impotence, detumescence, and cuckolding alike, and as further vestiges of the relation between theater and Priapo-Dionysian reveling. Indeed, Nancy, following Ferenczi, for whom a red forehead is equivalent to an erection, will argue that Descartes's whole embarrassment functions as a type of sexual-ontological disarray:

> The portrait of Descartes that no one has ever seen, and which however is his only portrait, is that of an erect Descartes. . . . The mask dissimulates the subject in disarray, in disorder [*en déroute*], ashamed of himself and bereft of assurance the moment he presents himself. But since the subject *was not* before coming out upon the stage where he will say "I am," his mask also dissimulates the confusion in which he comes unto himself (and thus unto being).[12]

With every erection comes a fear of falling or failing, and a pure physiology of sexuality is a first casualty of the thinking and psychologizing animal: henceforth no (sexual) performance without anxiety. Thus we can imagine the sort of disarray Nancy describes not just in the case of a coming to being such as will be Descartes's but also as a function more generally of our coming to be human. The anxiety would then extend beyond the question of tumescence to the upright stance as a general fall out of the animal and into the human, into "uprightness as erection in the general process of hominization" (*Animal*, 61), with its accompanying fear and shame of falling. The upright stance brings about not just a new and different exposure of the sexual organs but also a new experience of erection itself as constituting the human. That perhaps involves the shame (and the pride) of abandoning one's animal past, the shame (and the presumptuousness) of vertical or erectile ambition, of a contrived realignment of corporeal articulations, the Promethean hubris of a technologically enhanced biped refusing to accept the quadruped lot dealt by nature. Indeed, it means a new linearity involving a new relationship to gravity and a verticality that signifies uprightness as pride and probity, elevatedness and rectitude. I would argue that a whole new exposure, indeed, a whole new dimensionality, comes into effect in the upright stance, a *frontalization*, precisely, that will produce the face and hence the mask; and at the same time, just as importantly, if not more importantly, a *dorsalization*, the invention and production of a generalized back or behind. If that dorsalization is at the same time a technologization, it is not just because the hands are freed to manipulate tools but also because the animal, an animal, reengineers itself and intervenes in its own articulations, both interior (corporeal) and exterior (with animate and inanimate alike). The interior realignment of the spinal column and repositioning of the cranium necessitates very different relations with the earth, with other animals, and with objects. It brings its own risks, which means that such a transformation is accompanied by fears of reversion (the shame of a fall), of subversion (the limp of an inhibition), and of perversion (the contamination of a prosthesis). We fall or fail as we ascend less because an accident can befall us than because what we call progress never occurs in absolute terms but at the price of various compromises or sacrifices (for example, here, diminution of auditory and olfactory sensorial functions). The *ego conscensurus* is therefore as masked as the *ego larvatus*; it borrows the persona of the *homo erectus* and *zoon tekhnologikon*, tentatively at first, putting on the hairless face it doesn't yet know it has, and certainly doesn't know how to use, projecting overtly a sexuality where pride and shame both hide and seek, and exposing a whole new back to friend and enemy, fellow and object alike, new fear and new passivity, new adrenal zones and regionality.

The *Cogitationes Privatae* (almost certainly not Descartes's title) from which the *larvatus prodeo* jotting is drawn, are part of the *Opuscules of 1619–1621* that appear in volume 10 of the Adam and Tannery edition of the complete works. There are only a handful of pages of notes like it before they digress into something of a mathematical treatise, supposedly as a result of Descartes's exchanges with Isaac Beeckman. However, those few pages include his 1620 reference to the "fundamental principles of a wonderful discovery" (*Writings I*, 3) as well as mention of his famous dream of November 1619 (*Writings I*, 4). Some eighteen years will elapse before the publication of *Discourse on the Method,* which appeared anonymously in French, in Leiden, in 1637. The second part of the *Discourse* begins, of course, with an account of that earlier oneiric moment, which his biographer, Antoine Baillet, situates and dates close to Ulm, on November 10, 1619. Here is how Descartes sets the scene, in the *Discourse,* of dreaming and thinking alike: "At that time I was in Germany, where I had been called by the wars that are not yet ended there. While I was returning to the army from the coronation of the Emperor, the onset of winter detained me in quarters where, finding no conversation to divert me and fortunately having no cares or passions to trouble me, I stayed all day shut up alone in a stove-heated room, where I was completely free to converse with myself about my own thoughts" (*Writings I*, 116).[13] It is not until the fourth part of the *Discourse* that we are treated to the *ego cogito* that therefore is, whereas the dreams—for there were in fact three of them—will out, in immense detail however, only via Baillet's biography. Four years after the *Discourse,* Descartes will publish openly, in Latin, in Paris, *Meditations on First Philosophy,* the second of which famously advances the proposition "I am, I exist [*Ego sum, ego existo*]," even though "I do not yet have a sufficient understanding of what this 'I' is" (*Writings II*, 17).

There is no need to repeat here what has been exhaustively examined and debated regarding the confusion of identity between autobiographical and philosophical subjects in Descartes, beginning with mostly naive assumptions or conjectures—which may well have been initiated by Descartes himself—about the layers of pseudonymity or anonymity that constituted his masking and the real identity that would eventually be revealed underneath. Those assumptions and conjectures begin with the *larvatus prodeo,* which is invariably put into association with the fifth entry (according to the current accepted order) in the notebook, where a pseudonymous Polybius the Cosmopolitan announces a mathematical "thesaurus" that ends with a form of dedication to German Rosicrucians.[14] Nor shall I return down the well-trodden path that reads his dreams in relation to his philosophical hypotheses. Such discussion also begins with the later entries in the *Cogitationes Privatae* concerning the 1620 discovery and 1619 dream, just mentioned.[15] I shall take as read the idea that the prosthetic constitution of the "I" that we have examined in the 1619 note continues to function, albeit each time differently, through the abandoned *Rules,* to the anonymous *Discourse,* and into the fully self-conscious *Meditations.* And I shall leave open the question not just of the philosophical status, or the fictive status, but more particularly of the autobiographical status of a dream (recounted by a biographer).[16]

Instead, we might reapproach the question of autobiography from the perspective of "what lives" as it were prior to any published Cartesian *ego—larvatus, cogitans,* or otherwise—but still in order to loop back to the one, such as it is, that announces its first move in 1619. As Derrida puts it in *The Animal That Therefore I Am,* "however problematic it be . . . the characteristic of what lives, as traditionally conceived in opposition to the inorganic inertia of the purely physical-chemical," is "an auto-kinetic spontaneity . . . power to move spontaneously, to feel itself and to relate to itself," an "auto-affection or auto-motion" (94); elsewhere he calls it "sensibility, irritability, and auto-motricity, a spontaneity that is capable of movement, of organizing itself and affecting itself, marking, tracing, and affecting itself with traces of itself" (49). For that traditional conception, life begins with the spontaneous movement that Lamarck found to be the sole property of animals.[17] But, Derrida continues, "what is in dispute—and it is here that the functioning and the structure of the 'I' count so much, even where the word *I* is lacking—is the power to make reference to the self in deictic or autodeictic terms, the capability at least virtually to turn a finger toward oneself in order to say 'this is I'" (94). Supposedly only the human animal is capable of "this autodeictic or auto-referential self-distancing [*autotélie*]" (94). The distinction between whatever lives and the human animal would be determined in those terms as a difference between simple autokinesis, on one hand, and autobiographical autodeixis, on the other. Derrida's recognition of a problem or problematic within that traditional distinction derives from a number of factors that would make it difficult to determine with all necessary rigor what constitutes the minimal autodeictic or autobiographical "impulse." If, for example, an organism is able to replicate—which means "read"—its own vital codes and thereby self-(re)generate, does that constitute a sufficient case of self-recognition to be called the writing of one's life? If there were something in such an organism that motivated it to move in space and time, to consequently affect its environment and thereby constitute some form of traceable differential or dynamic existence, some type of becoming other, would that sufficiently transform its auto-affection into a hetero-affection that might be identified as auto-*deposition* or, as it were, the independent mark of a self? Much further down the evolutionary line, let us say, if an animal were able to recognize itself in a mirror, or recognize its species by means of the purely visual perception of an artificial model, which is the proven case for various species, even, as Lacan reminds us and Derrida discusses (*Animal*, 121–22), to the extent of their being sexually stimulated by such a simulacrum, would we not have to recognize there the formation of an autodeictic identity? What, then, is merely automatic or mechanistic tracing, and where does self-conscious life-writing begin? Where does the animal, in particular the human animal, break off from its biomechanical automatism to become an autodeictic "I"?

The answer to that question consistently involves recourse to another presumed distinction, that between reaction and response. Animals react, whereas humans respond. That distinction is Descartes's recourse, especially regarding the automatons described in part 5 of the *Discourse*, as discussed in chapter 9. On it, quite obviously, there hangs the whole ethico-ethological discourse of agency and responsibility, which presumes that only *respons*-ive humans can be *respons*-ible, but which fails to acknowledge the residual structure of automaticity functioning not only in reaction but also in response (cf. *Animal*, 125). For our purposes here, rather than take the question "forward" toward an ethics of responsible agency, could we instead ask whether, in its "inception" or "conception," automatism could be understood to function "beyond" or "before" animal reaction, as a technological necessity lodged in life itself? That would necessarily be so, Derrida argues, once the living being is defined in Lamarck's terms as possessing "automotive spontaneity" or the capacity to move by itself, for such a capacity cannot avoid a type of automaticity:

> the living being [*le vivant*] concentrates in a single ambiguous value this automotive spontaneity... [which] is right up close to automotive autonomy but also signifies its opposite, namely automaticity, or in other words the automat's mechanics of action and reaction.... The living being is automotive, autonomous, absolutely spontaneous, sovereignly automotive, and at the same time perfectly programmed like and automatic reflex. (*B&S I*, 221)

What is described there is not just the mechanics of physiology that Descartes described in great detail in his *Treatise on Man*, summarized in part 5 of the *Discourse*, nor simply the chemical reactivity of adrenal functions or synaptic operations, nor even just fear of being returned to deathly inanimateness by the encroachment and uncontrollability of proliferating technological apparatuses. It is rather the necessary structure of an originary mechanics, some form of inert inanimation at the very heart of the spontaneous capacity of the living to move by itself, some "cadaverous physico-chemical" (*Animal*, 49) in life, composing it. Life, in being or becoming life to the point of inscribing itself, would thus inscribe also a contradictory or paradoxical "force" of inanimate matter, a repetitivity or automaticity such as we would normally ascribe to what is contrary to life.

That tension, contradiction, or paradox can be seen to turn around the concept and functioning of autobiography. It is seen, to begin with, in *the writing of the life of a self* to the extent that that straightforward understanding of *auto-bio-graphy* is necessarily testamentary. One writes an autobiography in order, one way or another, to set the record straight in the perspective of one's death, the return of one's body to inanimate inorganicity and lifeless inertia. Indeed, as we know, as soon as one assumes a proper name, and even more so as soon as one *signs* that name, and even more so as soon as one signs that name on the outside of a piece of writing whose inside speaks of an "I" (or a "he" or "she") that is presumed to be in a relation of adequation to the name appearing on the outside, then one has been tagged with something that will function beyond the term of one's mortal existence. One's proper name is in this way both one's proper name and the name of one's very own inevitable death-to-come. What characterizes the autobiographical animal once it is able to point autodeictically to itself to the extent of having a proper name is thus the structural fact of death. An animal with a proper name is an animal with something that will survive its death. Its name signifies its death-to-come, and that supposedly inanimate name will continue to "live" once the animate animal itself has become cadaverously inanimate.

Now the proper name that *biographizes* the human is again something the animal is presumed to lack, except for those animals we anthropomorphize up to that particular point of domestication (or domesticate up to that point of anthropomorphization). Yet, even though the cases where I just suggested that automotion was perhaps already autodeixis—genetic reproduction, automotive heteroaffection, simulacral recognition—cannot be easily interpreted to mean that animals are also on the road to naming themselves, it would nevertheless seem clear that what we call naming is but one autodeictic instance, a particular form of identifying based on human conceptions of language. That is brought out through analysis of "our" primal mythological narrative of naming, the biblical scene where God calls on man to name the animals. God himself has and needs no proper name; he is indeed unpronounceable, pure autodeictic performance— performed as it were without motion of any sort—that gets uttered as the original pleonastic ego: "I am that I am" (Exodus 3:14). He is an *ego* that just is, a *sic ego,* without qualification or syntactic complement: *sic ego,* just that, nothing more. Man, conversely, has as yet no proper name; he is as yet man alone, male human only, and it falls to him to give common nouns as names to each and every animal: "So out of the ground the Lord God formed every animal of the field and every bird of the air, and brought them to the man to see what he would call them; and whatever the man called every living creature, that was its name" (Genesis 2:19). At that point in time, nameless man's naming of each and every animal can be understood from a variety of perspectives: first, the common noun he will invent for each animal is in effect the attribution of a proper name, because the very distinction between proper and common is not yet operative. Adam's own name, once it comes to be used a few chapters later (proper names really begin with Cain and Abel and the genealogies of Genesis 4), will have similarly derived straight from the word for the human species; it is simply the name of man (Heb. 'adam, formed from the humus ['adamah]).

Second, man's naming of the animals is his first recorded utterance, which means that he names the other(s) prior to any self-attribution whatsoever; and third, this appears as man's first technological moment, his first archival exteriorization or creation of an artificial memory, hence a stumbling into technology prior even to the Fall that is about to take place: man clothes the naked animals with a name before he even knows what nakedness is, in his or in anything else's regard. In the light of what I said earlier about the conceptual machinery required to have knowledge of nakedness as becoming-technology before the fact of, say, clothes, we would now have to hear that machinery starting up in the event of naming the animals. Consciousness of nakedness, of good and evil, of sexual difference, of self, and of all the rest would have to be understood to begin with the attribution of a name. The name is in that sense our first fig leaf or garment and therefore our first mask. Where there is naming, there is masking: man advances into life by holding out a mask in the form of a name, of his pure invention, for animals to put on. From that point on, everything he attributes to himself will come to him from the other, from the mask of the other, from nonknowledge regarding himself, from a type of self-imposed visor effect, which is also to say from behind, outside of what he can see or foresee.

That would be, in terms I have argued elsewhere,[18] the case of technology in general: first, it is not an invention of the human, in the sense of an invention by humans, without also being the means by which the human itself is invented, and that takes place precisely through and as invention itself. On God's invitation, man acts like a god of creation, conjuring up, as if out of nothing, names for the animals. Not quite able to repeat the gesture of an "I am that I am" and forgo a name for himself, he instead declares, by means of his performance, "I am he who invents names; I, who as yet has no name, will make a name for myself as the inventor of names." His inventions will have a life of their own (the names, rather than the animals bearing those names) beyond the moment wherein he creates them, and even beyond what he can foresee of his own finite existence. The invention is performed within that very perspective; that is its sense as performance. As it were superseding the name of him who comes from the earth, but before the name with that sense is capitalized to become properly his, there is for the human this name of "namer" or inventor; he is that which is constituted by the capacity to invent.

Yet that all comes to him *(invenire)* from outside of his perspective, as it were, preceding him; for the performance is directed by God, with the animals as its actors ("[God] brought them to the man to see what he would call them"). Not only is the privilege or responsibility of naming the animals not something that man is expecting but the names themselves, however much he invents them in repeated strokes of creative nominal brilliance, have to be owed to the animals themselves, as if each of them came to him to say, "Call me this"; even their simple presence, the fact of their stepping forward one by one, collaborates in the process. God is there watching to see what man will call each animal, and each time we have to presume that man looks at the animal and looks to the animal (or looks to God, or anywhere else—it amounts to the same thing) for inspiration. The names don't preexist in man, for no experience of his or of the world has prepared him for this; he has no experience, at least not of the animal world. So he comes up with a name thanks—at least in structural part—to the animal that stands before him, yet the name comes as it were from behind him, or at least from outside his experiential field, as though it were out there waiting for him to find but at the same time within him.

In Benjamin's famous analysis of this biblical narrative, "On Language as Such and the Language of Man," to which I return in chapter 5, the naming of animals functions within the context of a more general naming of the world. For he contrasts the naming that defines the human relation to language—"we . . . know of no *naming* language other than that of man"—with not just animal life but also "inanimate nature." That is something of an absolute contrast, and it leaves nature mute and mourning, precisely for having been named and thus becoming "thoroughly known by [what is] unknowable."[19] Hence, according to that view, naming is a sovereign act performed on a passive entity whose status as object is correlatively intensified—the obedient animal (or nature) is called forth, given a name, and rendered all the more passive, mute, and mournful. But my hypothesis, that the name comes to man from the other that he names, still holds, I contend, once the operation is extended from the animals explicitly referred to in Genesis to nature as a whole. At the beginning of Benjamin's article, it is said to be "in the nature" of each "event or thing in either animate or inanimate nature [to] in some way partake of language" (*Writings 1*, 62), which allows us to understand the perceptible world in its entirety as communicating its linguistic being before or beyond any specificity of the linguistic being of man defined in terms of naming. Man may invent names, even the language of naming, but he does not invent language itself; even mute nature, from this point of view, utters, even if only from the impotence of its lamenting. That means that the world as a whole is technologized by language prior to being convoked specifically by human naming, so that the name that is "given" to the human by each "creature"—animate or inanimate—in the grand nominational ceremony is, as much as a word for what it will be called, the name "technology."

Man's act of naming, therefore, should be understood as something other than a sovereign act of autonomous deixis, one that would define the animal as what is named and, conversely, define the human as the operator of naming, capable of naming others because he understands his own name and status. The biblical narrative describes, rather, a moment of structural heteronomy: man is *called upon,* by God on one side and the world on the other, as much as he *calls*—and not just *called upon,* as if by other objects to which he will have to relate as subject or object, but *called out of himself,* into a technological articulation that is before and thus behind him. His inventions, before being represented as this or that name, are an effect of that articulation, which binds him prosthetically to a whole network of othernesses that require him also—and this is strongly reinforced by Benjamin's reading—to redefine his ethical relations.[20] They require him to understand the other not just as what is "facially" opposite but also as what challenges from the fringes and outside the range of conceptual possibility even up to the point of inanimate irreconcilability.

In a profound way, therefore, at the moment when man is called on to exercise his sovereign authority by naming the animals, he abdicates his autobiographical autonomy, or at least autobiographical priority. Whether we understand that in the simple terms of narrative sequentiality or in the heteronomous structures that I have just described, it is the animals that end up being inscribed with autobiographical potential: either by virtue of being named first, or by virtue of stepping up to man as if with their names. When God brought the animals to man, man could have said, "Hold it, God, I'd like to be sure about my own name and self first," but he didn't. His first performance, therefore, had its performativity voided, as it were; he spoke and acted not by uttering his self but by conferring a nominal selfhood on the other. Now, one might still insist on interpreting the scene as a purely arrogant performance of sovereign power, man's uttering and imposing his dominion over the animals—imposing on them his law of the name of the father—exercising unilaterally the power that is expressly conferred on him in the first version of the creation narrative of Genesis 1 and that will be holocaustically confirmed, soon after this second version, by means of animal husbandry and sacrifice. That would still be correct. But in this scene, there is only God on one side and the animals on the other to witness the performance of man's dominion. It is not clearly the scene of a stage in Descartes's sense, for there is, strictly speaking, no biographical record of it other than the fact of the names themselves: no one endowed with a proper name is there to witness it, neither God (whose name cannot be uttered) nor man himself (still 'adam, not yet Adam), nor the animals (given names that are indifferently common and proper); and clearly there is no witnessing in the sense we give to the word before there is a proper name.

So man is seen to falter in his first autobiographical moment. He goes forth expecting to mount the stage with his own name and instead finds himself masked by the names of myriad animals. Instead of giving himself a proper name that would enable him to act in his own name, he confers a type of selfhood, in multiple or infinite form, elsewhere. He thinks, perhaps, that he is making a preemptive strike against death, that by taking on the task of naming every other species, he will be able to stave off his own demise, defer the moment when he will have to give himself a name and so sign the mortality register. And as I noted earlier, he never actually did that for himself, nor did God. Man simply adopted the common noun of his species as his proper name, just sidled into it at some point, quite possibly around the very time he was naming the animals and waiting for the creation of woman, a woman whom he would, to be sure, name: Eve, the mother of all that lives. He therefore defers his own genealogical and autobiographical mortality by setting in train the endless naming machine, and the story of his life begins by being, and is structurally constituted by being, this impersonal imprinting of another's self. His is therefore an ego that asserts itself, and so undermines itself, by first giving selfhood to the other. It may well be giving death or submission to the other, and life in the sense of power to exploit the other, to itself. But it is also giving death to itself by instituting the archive of its own life as the recording of every other life. At this point in history, man's life story consists of an interminable list of names concerning everything but man. We can't even say that each name of each animal expresses and records his will or his whim, for he hasn't yet established, for the public record, an *I* that could will, and God alone knows what sort of knowledge he could draw upon to form the content of that will or whim. He has simply started up a biographical machine without an *autos,* but one that will run interminably, as if automatically.

So man dies a little, a little death, each time he utters "I," and that dying begins with his deferring of uttering his own "I" in favor of naming each and every animal, which makes man heterodeictic before he is autodeictic and confers upon the nonhuman a prior autodeictic capacity while constituting deixis itself as machinic, automatic, inanimate. Deixis itself becomes automation not just because, in the narrative, there were an infinite number of species to name but because of the simple fact of repetition. Man could have broken off after a while and said "now the rest of you figure out your names for yourselves," but he couldn't have named fewer than two animals. The name of a single animal, common or proper, would mean nothing by itself. The name, first of all, would have to have differential effect vis-à-vis another name, even if that other name were namelessness. But more importantly, the deixis itself would not be able to function, and neither man nor animal would know what was going on, before the deixis had repeated and cited itself. A deixis or pointing out that takes place once only has no more effective sense or value than a signature signed once only.

So man gets going pointing and naming, and never stops, even if he gets to the end of all the species, for the fact of pointing that he has instituted continues to function without him. He may have thought he was keeping himself immortal by deferring his autobiographical moment, but that immortality was deriving no more from some superhuman energy he mustered to fulfill to the very end the task God had given him than from a self-replicating code that began the moment he pointed for the second time, a code that would necessarily still be in operation, structuring the very thing, once, if ever, he were to get around to pointing to himself.

Yet perhaps that is precisely what we should imagine life to be, beyond any autokinetic–autodeictic distinction; or we should imagine life to be the *automation* of *automotion,* the autokinesis at work in any autodeixis whatsoever. We can observe it, for example, in autobiography and understand autobiography as something other than the writing of one's life in the prospective of death—something other than simply what survives the end of a given life. It is not just the inanimate writing that comes to life to prolong a life once the human body that previously represented that life returns to dust; instead, it is something like a graphic automation or inanimation that precedes and even gives rise to life. Any traceable self, any replicable *autos* whatsoever, might be understood in this way to write itself, its life, automatically, from the beginning, and autobiography as the archiving of one's life would be coextensive with what we call natural, spontaneous self-generation or reproduction. For the logic of autobiography is that of a type of automatic regeneration: as I record my life, I add to the life that my autobiography will henceforth have to take account of, along a future vanishing point that only death can interrupt. A life comes to be constituted by everything lived plus the writing of it, ad infinitum. I have done this, this and this, and written about this, this, and this, which means I have done this, this, this, and this and now have to write about this, this, this, and this, and so on. From that point of view, autobiography would be something that extends life in the sense of producing life, and we would understand that function of autobiography not as something accidental or contingent to it but rather as its constitutive structure. Autobiography is the name of what, in life, allows it to be automatically extended. Which also perhaps means that the concept of autodeictic autobiography is *necessary* for there to be life—that what keeps me alive is my saying *I,* and life exists, life continues, life is produced to the extent that I, an animal, some thing, am able to do that, to say *I.*

Sic ego, writes Descartes, in the inaugural moment of his writing self. Having written it once, he will, as we know, spend much of his life rewriting it. His life, and afterlife, will be defined by it, even as he thinks that he has definitively defined it. *Sic ego,* so it is with me, so I (too). So, it is as if he writes, this *I* will be the question of my life; it is what I'll put on stage, starting now, and ending definitively by starting definitively the modern philosophical moment. This *I* will be my signature, beginning now and culminating in the *ego cogito. Sic ego, here am I,* he wants to say, like Abraham, replying self-assuredly to the call of God's angel at the moment of raising his knife to sacrifice Isaac, the knife that will assuredly fall henceforth instead on animalkind. He wants this to be his Abrahamic moment, inaugurating a long and solid tradition. But, instead, his first self-writing moment will be his Adamic, or '*adamic,* moment. He begins by naming the others—actors who put on a mask—and the I once named will never truly be his own, never nakedly his. In fact, he also calls the animals. Some pages later, in the last properly philosophical fragment of the *Cogitationes Privatae,* before it becomes mathematical, he writes that "the high degree of perfection displayed in some of their actions makes us suspect that animals do not have free will" (*Writings I, 5*). Animals are pure actors, such perfect actors that they are only actors, almost automatons, unlike human actors, who decide to be actors. Like me. It is like this with me: I name the others actors (including all the animals) so as all the better now to point to myself, an actor, saying this is I, thus I (am). I am not about to be conned, by God or any other animal, into an interminable naming that preempts my autobiographical possibility. Descartes dictates a simple differential series: there masked actors and unmasked animal actors, here I, and beyond me the I that will don a mask and advance. But at the point he writes *sic ego,* he is not quite there. He is not yet, not quite, pointing autodeictically to a becoming autobiographical self. When he writes "there actors . . . so I," he is not yet writing "I am an actor, one who will advance masked," however much that is implied, as I began by strongly arguing. *Sic ego* is not quite the same as *hic ego* (here I am). That is because the very *sic* precedes and interrupts the *ego* even as it positions it and syntactically aligns it as the subject of the verb *prodeo.* The *sic* is the precise point on which it all turns, and it turns backward and forward at the same time. *Sic,* thus, so, in that way, as I was saying back there just now, like actors, as it was with them there, so it is with me here. *Sic* says "the way it is with those actors is what I am now going to apply to this *I*, to my own case." But before being able to operate as the forward force—perhaps eventually the will—of an I that will advance onto the stage of the world and of philosophy, it will have to reach back to and reckon with what it owes to what is behind it, there where some more originary automotive (im)pulsion is in operation; there where there is *acting*. Acting before actors. Without such acting, the automotion of life, there are no actors and no I. *Sic* here carries all that, it is that around which all that turns. Before any *I* comes to be, *it is the thus that is*: the *thus (sic)* that is and must be before any *therefore (ergo)*, before any thinking I therefore is. *Sic* is thus the movement of Descartes's arm as it pivots from pointing to actors and flexes into autodeixis. It *is* that very pivot. It is the articulation of an elbow and carries with it articulation itself, a simple mechanics of movement that allows every thing to join itself to every other thing, the point and operation of a generalized prosthesis. So *sic* works out there before the *ego* can utter, affirm, or define itself; it is like a technology preceding it. And *sic* precedes the ego in the sense of being behind it, not just because it got there, got out, first; more importantly, it operates behind the ego because it is *what is necessary* for the ego to begin to be anything whatsoever, for autodeictic life, life in general, to begin. It is the automatism of the motion of life, what moves in living as movement, what lives in moving as life, the very redundancy or tautology of that. Perhaps life itself, even if not itself life: inanimated life, in that manner, way, form.

So, life.

2
ORDER CATASTROPHICALLY UNKNOWN
Freud

Suppose such a syntagm: "Order catastrophically unknown." Flashed on a screen in some operations room crowded with aghast personnel and overheating machines alike: the personnel either brow-furrowed Americans in crisp uniforms or incomprehensible transgalactic aliens, or a combination of one and the other; the machines a serial constellation of flashing lights, pixelated frames, maps and increasingly detailed GPS zooms, dials, levers, and buttons. Flashed on a screen or announced in staccato monotone by a synthesized android voice-over name of Hal or Ethel: "Order catastrophically unknown." Imagine any version of that from any film or TV series that one may or may not avow knowing or loving—*Lost in Space, Dr. Strangelove, Battlestar Galactica, The Bourne Ultimatum*—any of those of the last forty or fifty years, the time of our current technologico-revolutionary suspension. Or imagine it emitting from a time period more or less twice that length, the time, say, that is simultaneously triplicated as the long twentieth century, the time of cinema and/or photography, and the time of psychoanalysis. "Order catastrophically unknown." Suppose it, in that context, as the syntagm, something like the first utterance regarding what we now presume to recognize as the unconscious. Some previously undiscovered and unclassified life-form, just emerging into taxonomic space. Some animal like that. A first man called Freud stands forth, like Adam in Genesis 2, under the gaze of God, to respond to the call of naming the creatures that parade before him. He begins with eel gonads in 1876 in Trieste, moves on to lamprey larvae, and then, twenty-odd years later, spots a strange new species he will call the unconscious. A chimera, hitherto unnoticed, or at least un-named. Suppose said syntagm as the taxonomic quandary of being faced with such a new life-form. Or suppose it, finally—"order catastrophically unknown"—as *that very unconscious itself* uttering itself, perhaps in the guise of the death drive or the very inorganic origin that gives rise to it. Something that couldn't be, doesn't know how to be, is no longer once its uttering takes the form of a coherent syntagmatic chain articulated by means of coded linguistic forms employed by a rational living being.

The syntagm I am asking to be supposed is a paraphrase of a quote from "Envois," in Derrida's *Post Card*: "My post card naively overturns everything. In any event, it allegorizes what is catastrophically unknown about order" (21).[1] The French is "l'insu catastrophique de l'ordre," which might be rendered more literally as "the catastrophic unknown concerning order." The order Derrida is referring to is, in the first place, *sequential* ordering. He continues, "Finally one begins no longer to understand what to come, to come before, to come after, to foresee, to come back all mean" (21). But one should also read it, in the context of the generic or taxonomic conundrum that Derrida wants his postcard to represent, as the catastrophe of what is unknown concerning *classification*. In the case that I want to concentrate on here, that of lifedeath, animate–inanimate relations, both orders are in question: what comes first, and how something is to be classified. As Derrida states in the opening of "To Speculate—on 'Freud,'" originally part of his 1974 "Life Death" seminar, "the issue . . . is *to rebind . . .* the question of *life death* to the question of the *position* [*Setzung*], the question of positionality in general, of positional (oppositional or juxtapositional) logic."[2] Psychoanalysis, he will persistently argue, with its delayed effects, returns and reversals, overturns the simple linearity of causal relations. And the Freudian unconscious is, it seems hard to deny, in constant morphological and categorial flux.

This chapter will trace a certain—perhaps fractal—outline of those morphological shifts, particularly in the metapsychological works beginning around 1915 and culminating in *The Ego and the Id* of 1923, and consider the consequences of such transformations for the question of life, a definition of life overturned or undermined by the death drive. It will cut across the broader question of the organicism of psychic life that Freud poses from one end of his work to the other and whose ambiguities come to be concentrated in these works of retheorization from his middle period. In "On Narcissism: An Introduction," from 1914, he claims that he tries "in general to keep psychology clear from everything that is different in nature from it, even biological lines of thought." However, he is required to add, "the hypothesis of separate ego-instincts and sexual instincts rests scarcely at all upon a psychological basis, but derives its principal support from biology."[3] Similarly, whereas in the same text he writes that "all our provisional ideas in psychology will presumably some day be based on an organic structure," which "makes it probable that it is special substances and chemical processes which perform the operations of sexuality" (*SE XIV*, 71), he will argue instead a year later, in "The Unconscious," that the physical characteristics of the latent states of mental life remain "inaccessible" and that "no physiological concept or chemical process can give us any notion of their nature" (*SE XIV*, 168).

The question, and its attendant ambiguities, have been well treated of, generally speaking, as a tension between his initial background in the biomedical sciences and his more or less independent development of what he will want to call the science of psychoanalysis. The presumption is that in developing, or redeveloping, a theory of the instincts, which, "from a *biological* point of view," appear as "a concept on the frontier between the mental and the somatic" (*SE XIV*, 121–22), he was required to incline now to one side, now to the other of that frontier. What is perhaps less well treated of—hence the importance of Derrida's emphases—is, in the first place, the stress that the discovery of the unconscious brings to bear on the mechanist causal principles that governed science in general, which might also account, but differently, for the constant modifications of psychic topography, systematicity, and dynamism that are advanced in the metapsychological texts.[4] In the second place, as I argue here, that stress leads as far as a catastrophic unknown concerning not only the order of cause and effect but more particularly the order of life itself.

By 1920, the border between soma and psyche will devolve upon that between life and death. Though the study of the sources of instincts was supposed, in 1915, to lie "outside the scope of psychology" (*SE XIV*, 123), the logic of a beyond the pleasure principle will lead, in the text of that name, to an investigation of "forces [*Kräfte*]" operating in "living substance" but nevertheless held to be distinct from it (*SE XVIII*, 46). However, once the instincts are understood to function in competition one with the other, once they are defined as those of life and death, and once death is conceived of as a return to a prior, preanimated state, then the distinction between living substance and instinctual force within it will become a problem whose resolution will necessarily involve something of a search for a source or origin of life. The instincts or forces within what lives will progressively be found to be as originary as living substance itself, indeed, as a form of life within that living substance, even if it has the form of *lifedeath. The Ego and the Id* repeats, in 1923, the conclusion effectively reached in *Beyond the Pleasure Principle* regarding the instincts' animation of living matter down to its very origin. Both life and death instincts, Freud writes in the former volume, "would be active in every particle of living substance" (*SE XIX,* 41).

By the end of *Beyond the Pleasure Principle,* Freud still seems to hope for a scientific solution to that problem of defining the active force of the drives within living substance once the particles of living substance itself are traced back to their chemically active, but inorganic, origin, and once one of the drives within them is understood to involve a return to the inanimate. The solution might be derived from a simpler, chemical language or borrowed from the "land of [truly] unlimited possibilities" called biology. Indeed, it is said, the answers provided by biology "may be of a kind which will blow away the whole of our artificial structure of hypotheses [*die unser ganzer künstlicher Bau von Hypothesen*]" (*SE XVIII,* 60). Biology, in that formulation, would function as a force of living truth capable of overwhelming the inanimate construction of a theoretical apparatus however complex, capable of replacing the prosthesis of metapsychological suppositions with scientific fact. In the wake of such a biological possibility, there would be no more death drive, perhaps no more psychoanalysis. Conversely, however, as long as there is the type of speculation that leads beyond the pleasure principle, psychoanalysis will continue to function as something of a death drive within the science of life itself, returning it to the artificial mechanist framework of its pre-eighteenth-century formulations or insisting that what animates it still is both an organic logicality and an interruptive and contrived hypothesizing.

The catastrophe of the unknown concerning the order of life, its catastrophic overturning, is therefore derived explicitly from the death drive. All the doubts and questions Freud expressed concerning his theory in the years 1895–1920 may be interpreted as symptomatic of the incongruous insight that suddenly flashes to consciousness beyond the pleasure principle, like a whole new unconscious, the unconscious of the unconscious, namely, the insight concerning the inorganic origin of life itself. Life exists, Freud explicitly writes in *Beyond the Pleasure Principle*, thanks to the fact that certain attributes "were at some time evoked in inanimate matter by the action of a force of whose nature we can form no conception" (*SE XVIII*, 38). What is on one hand a biopoetic or abiogenetic commonplace, namely, that organic matter derives from inorganic matter—for example, that "innocent" chemical processes at some point gave rise to nucleic acids—takes on a profoundly radical sense once it comes to be endowed with the paradoxical form of "intentionality" that Freud gives it, and more than that, with the reverse or conservative intentionality whereby "an instinct is an urge inherent in organic life to restore an earlier state of things" (36).

All that suggests that in the dim distant past of psychic life, there would have been—emerging from who knows where, but formed at the very moment of life itself, within the very structure of self-preservation—a proto-desire not to live. As if some 3 billion years ago there was a single-cell prokaryote that formed, took a quick look around, and said to itself, "Self, listen, I'll do this if I have to, but I want it on the record that I was happier before, simmering in the primordial soup." In that utterance, were we able to constitute it with any scientific or historical accuracy, and once we were beyond the pleasure principle, there would be contained the origin of consciousness and unconscious alike, and, what is more, after Derrida after Freud, a type of originary autobiography. For the further catastrophic overturning that takes place in psychoanalysis, and that relates to Freud's scientificity as much as to the lifedeath of the unconscious, is what Derrida refers to as Freud's business, in *Beyond the Pleasure Principle*, of producing the institution of his domestic and autobiographical desire ("Speculate," 299). The concerns of this chapter, following what was introduced in the previous one, will stand in relief to those general and specific considerations. It will be a matter of continuing to ask what life is, or lifedeath, given the unconscious and the death drive, and whether that question can be posed (after Freud, after Freud after Derrida, or ever) outside what we have already understood as another type of instinctual reinscription called autobiography.

In chapter 1, I related what Derrida calls "autobiographical auto-deixis" to Descartes's ego, precisely the proto-larval *ego larvatus prodeo* of his early jottings that would lead to the *ego cogito* of the *Discourse* and *Meditations*. Here, while respecting the differences, I should like to consider the ego that emerges, at the other end of the modern history of consciousness, from Freud's metapsychological writings, notably, of course, in *The Ego and the Id (Das Ich und das Es)*. *Das Ich* is not of course, in 1923, a simple transliteration of the *ego* that appears, through the accident of Descartes's Latin, in the formula "ego sum, ego existo . . . [ego] sum . . . res . . . cogitans" of 1641. Yet, on the other hand, Freud in *The Ego and the Id* is as anxious as Descartes was in his own way (via dreaming, evil demon, and *res cogitans*) to isolate the operative psychic configuration of the unconscious vis-à-vis types of consciousness, to determine what one and the other category is and isn't, their topographical, systematic, or dynamic relations. Consciously or unconsciously, it is not for nothing that Freud decides to call what we call the "ego" *das Ich* (and not, for example, *das Selbst,* which he uses in apposition to *Ich* on one occasion in *Civilization and Its Discontents*).[5] And so the accident is more on the side of the English translation that opts for the word *ego* rather than the *I,* thereby avoiding or repressing whatever ambiguities inevitably arise between the concept of the I *(das Ich)* and the I *(Ich)* that utters itself and signs itself as the author of the texts wherein that concept is developed. Yet "The I and the It" would be a more faithful translation of Freud's title, however much pronominal confusion might ensue. As we shall see, such confusion can be identified not just where a Freudian "I" advances a hypothesis concerning the libidinal ego's articulations with the id and superego but also where a type of larval life takes a first halting step in the name of self-exteriorization.

We know that as early as in his 1895 *Project for a Scientific Psychology,* Freud's intention was to "furnish a psychology that shall be a natural science" (*SE I,* 295). To that end, he developed the principle of "neuronal inertia," namely, that neurons tend to a state of rest, with the resulting hypothesis that the organism will withdraw from external stimuli by means of motor functions, or flight, but will be required to deal otherwise with endogenous (internal) stimuli such as hunger, respiration, and sexuality, flight not being an option in such cases (295–96). Chapter 7 of *The*

Interpretation of Dreams repeats the hypothesis: "at first the apparatus aimed to maintain itself as far as possible without stimulus, and so in its earliest structure it assumed the scheme of the reflexive apparatus, which enabled it to discharge in motor activity any sensory excitation reaching it from the outside. But this simple function is disturbed by the exigencies of life; it is to these that the apparatus owes the impulsion to further development" (*SE V*, 565). Something Freud calls the "exigencies of life" will therefore be the basis of psychic life, and the further development he mentions will be that of a psyche divided between perception–consciousness and preconscious–unconscious, under the direction of the pleasure principle. As is well known, that was the configuration of Freud's first topographical or structural model. As is equally well known, the second model would wait some fifteen to twenty years before it found its full exposition in *The Ego and the Id*. It might be said to begin with the distinction between ego-libido and object-libido that is made in "On Narcissism." Following "On Narcissism," Freud wrote twelve metapsychological papers, five of which have survived. In those papers, various modifications are made to the first structural model of the psyche, especially by means of attention to the instincts or drives as psychical representatives of "the stimuli originating from within the organism and reaching the mind" (*SE XIV*, 122). The theory of instincts will make the antithesis of conscious and unconscious no longer applicable: "An instinct can never become an object of consciousness—only the idea that represents the instinct can. Even in the unconscious, moreover, an instinct cannot be represented other than by an idea" (177). At the same time, various ambiguities are introduced, which are presumed resolved with the explicitation of the death drive in *Beyond the Pleasure Principle*. For example, the ego itself, though referred to via the ego-libido and ego-ideal, remains undefined, and the object-libido–ego-libido distinction remains unclear. Indeed, more generally speaking, apart from identifying reality testing as "among the major *institutions of the ego*," Freud freely admits in "A Metapsychological Supplement to the Theory of Dreams" that "we know too little as yet of the nature and mode of operation of the system Cs" (233). It has been suggested that those ambiguities were discussed in one of the seven missing, lost, or disappeared metapsychology papers, particularly that on "the Conscious," but the absence of that paper also demonstrates the extent of

Freud's dissatisfaction with whatever he formulated in it.[6] As of 1917 and "Mourning and Melancholia," or perhaps as of 1920 and the time beyond the pleasure principle, we are left with an ego constituted by three major institutions: reality testing, censorship, and conscience (which will come to be called the superego) (*SE XIV*, 247).

Throughout the metapsychological papers, however, the now twenty-year-old principle of the *Project for a Scientific Psychology* is repeated with some insistence, and without modification: "We see then how greatly the simple pattern of physiological reflex is complicated by the introduction of the instincts. External stimuli impose only the single task of withdrawing from them; this is accomplished by muscular movements. . . . Instinctual stimuli, which originate from within the organism, cannot be dealt with by this mechanism. Thus they make far higher demands on the nervous system" (*SE XIV*, 120), or "One of the vicissitudes an instinctual impulse may undergo is to meet with resistances which seek to make it inoperative. . . . If what was in question was the operation of an external stimulus, the appropriate method to adopt would obviously be flight; with an instinct, flight is of no avail, for the ego cannot escape from itself" (146).

In introducing the distinction between ego-libido and object-libido in "On Narcissism," Freud has recourse to a telling biological analogy: "we form the idea of there being an original libidinal cathexis of the ego, from which some is given later to objects, but which fundamentally persists and is related to the object-cathexes much as the body of an amoeba is related to the pseudopodia which it puts out" (75). The ways of pseudopodia, Strachey tells us in an appendix to *The Ego and the Id,* are a favorite analogy of Freud's, being referred to at least four times: in the passage from "On Narcissism" (1914) just quoted; in 1916, in "A Difficulty in the Path of Psycho-Analysis"; in lecture 26 of the *Introductory Lectures* (1916–17); and, finally, in the *Outline of Psycho-Analysis* from 1938. Strachey's concern is less with the amoeba analogy than with ambiguities that Freud seems to introduce by means of references to "the great reservoir of the libido" (*SE XIX*, 30n1), ambiguities that confuse the distinction between ego and id. Strachey explains the confusion by asking us to understand the reservoir both as a storage tank and as a water supply and by admitting that Freud does not always distinguish as precisely as he might the topologies of an originary undifferentiated ego-id. Indeed, in lecture 32 of the *New*

Introductory Lectures, Freud states that "th[e] ego is the better organized part of the id, with its face turned toward reality. We must not exaggerate the separation between the two of them too much" (*SE XXII*, 93), echoing closely what he had already written in *The Ego and the Id*: "One must not take the difference between ego and id in too hard-and-fast a sense, nor forget that the ego is a specially differentiated part of the id" (*SE XIX*, 38).

I am not concerned here that Freud may have continued to wrestle with his second psychic model even after seeming to settle into his formulation of it in *Beyond the Pleasure Principle* and *The Ego and the Id* and beyond. Explicit references to the tentative nature of the id–ego–superego model are relatively common, as numerous as comments regarding the possibility, or not, of psychoanalysis's independence from biology or zoology. We can presume that as Freud developed his theoretical metapsychology, including the death drive, he became increasingly sensitive to how far he was departing from the scientifically logical norms that he nevertheless sought to respect. A scientific psychology remains his aim in lecture 32 of the *New Introductory Lectures*, for example, where he states, "Actually what we are talking now is biological psychology, we are studying the psychical accompaniments of biological processes" (*SE XXII*, 95–96). Thus biological psychology can be understood to function as something of a corollary, or counterpoint, to an originally undifferentiated ego-id, Freud's great theoretical reservoir.

The amoeba analogy is symptomatic of that. This unicellular protozoon, however different from Jacob's preferred bacterium by reason of its shape and reproductive method (mitosis rather than fission), is nothing more than a microscopic bundle of passive syntheses, a rudimentary larval subject in Deleuze's terms.[7] It nevertheless performs something noteworthy enough for Freud to raise the whole question of the psychic formations of nonhuman animals and, as I shall argue, much beyond. The most detailed exposition of it is found in the *Introductory Lectures*:

> To put the matter shortly, we pictured the relation of ego-libido to object-libido in a way which I can make plain to you by an analogy from zoology. Think of those simplest of living organisms [the amoebas] which consist of a little-differentiated globule of protoplasmic substance. They put out protrusions, known as pseudopodia, into which they cause the

substance of their body to flow over. They are able, however, to withdraw the protrusions once more and form themselves again into a globule. We compare the putting out of these protrusions, then, to the emission of libido on to objects while the main mass of libido can remain in the ego; and we suppose that in normal circumstances ego-libido can be transformed unhindered into object-libido and that this can once more be taken back into the ego. (*SE XVI*, 416)

As I read it, the pseudopodium introduces psychic operations into a most basic form of animal life. Freud probably doesn't want to suggest that the amoeba has an id or an ego—"today, our critical judgment is already in doubt on the question of consciousness in animals," he writes in "The 'Unconscious'" (*SE XIV*, 169)—even though he makes explicit, in *The Ego and the Id,* that "the differentiation between ego and id must be attributed not only to primitive man but even to much simpler organisms, for it is the inevitable expression of the influence of the external world" (*SE XIX*, 38). That basic response to stimuli, which Freud consistently believed to produce biopsychic operations, is what is being staged in the case of the pseudopodium. For science informs us that what motivates an amoeba to put out a pseudopodium, to create a temporary cytoplasmic projection of its cell wall, is a "desire" either to move or to capture food, or both. In Freud's terms, therefore, it is either a motor function response to an external stimulus, in the form of an attempt at flight from that stimulus, or else it is a response to an internal stimulus, namely, hunger; or it is both, indistinguishably exogenous and endogenous in its operation. But more than that, the pseudopodium is itself indistinguishably internal and external; to paraphrase Derrida, it is something of a *chiasmatic ex-vagination of the edges.*[8] It protrudes or projects from the cell and so reconfigures it, creating a new external surface and some form of independent appendage, which effectively, or operatively, divides the cell into main body and single limb. Then the amoeba reforms, cannibalizing or incorporating, perhaps introjecting itself, in any case reintegrating its previously externalized part—the pseudopodium—back into its whole.

On the scientific face of it, the pseudopodium exists only temporarily and should not be understood as a structural metamorphosis, even less so as a move toward cellular division and reproduction on the way, ultimately, to the sexual reproduction that was a necessary reference point for Freud's theory of the drives. But even Jacob conceded, at the level of the bacterial cell, the aberrant possibility of transferring "genetic material from one cell to another . . . either by the intermediary of a virus, or by processes *recalling the sexuality of higher organisms*" (*Logic*, 291, emphasis added), as if there were always the possibility of engaging with another to the extent of impugning one's own integrity, as if there were always that type of pseudopodial will-to-supplementarization—as if that sort of auto-hetero-affection were the model for automobile life in general, the impulse driving what Derrida, in reference to Jacob's observation, calls the "auto-reproductive autoaffection of the most obstinate bacterium" ("Life Death," 4:18). Even the most regularly formed nongroping cell will, it seems, reach out to the other, even if that other be the other within itself.

In that sense, the pseudopodium does operate within the structure, if not specifically in the direction or perspective, of cellular division and reproduction; it represents at least a tendency toward such division, comparable to it in morphological terms, a testing of those waters from which it then perhaps retreats. And even though in strictly causal terms the pseudopodium appears to serve a wholly different function from cellular division and is simply a reaction to a stimulus unrelated to DNA replication, Freud's analogical recourse to it explicitly raises the whole question of the origin and operations of endogenous stimuli, or instincts, and their relation to the preservation of the species, and that necessarily implicates in turn cellular, and then sexual, reproduction as a fundamental element in aid of such preservation. For the opposition, or imbrication of the instincts—self-preservative versus conservative—remains an explicit quandary of *Beyond the Pleasure Principle*, as indicated by the long note Freud adds at the end of chapter 6, attempting to explain how sexual instincts, less closely connected with reproduction, function as life instincts in opposition to the death drive, despite the libidinal nature of both ego instincts and object instincts (*SE XVIII*, 60–61). And, as if echoing Jacob, he has already raised the possibility "that the instincts which were later to be described as sexual may have been in operation from the very first" (41)—from the very first pseudopodial impulse, we may therefore surmise.

The *pseudopodium* is literally a false or fake foot rather than a *proto-podium* or rudimentary limb, or the *alethopodium* that is my prerogative here. I might, however, be forgiven for calling it a *technopodium* and recognizing in it a form of the originary technoprostheticity of the organism, *the fact of its always already articulating itself within itself in order to reach beyond itself.* Such an articulational possibility would be necessary for any relation to develop between a supposed intact integral entity and whatever constitutes its outside. It would not presuppose that relation to derive from either an external or an internal stimulus. Simply, relationality in general relies on the concept of an internal body or entity of some form articulating with the external space of whatever is understood to be foreign to it (another body, part of a body, whatever); yet for there to be that relation, or once there is that relation, the presumed internal body must also be understood to be structured by articulation, not just as future potential but as originary necessity. Every body—unless it is to die alone, never having the life of any interaction whatsoever—must harbor within it what might be called its *articulationality,* the capacity to alter its form in preparation for, or as a result of contact with, another body, where that formal alteration is understood not to take place on the body's exterior surface without also reconfiguring, bending, or articulating its interior. Thus I contend that there exists in the constitution of every cell the capacity both to change shape and eventually self-divide—a *morphologicality*—as well as an *articulationality* that does not structurally distinguish between the protrusion of a pseudopodium and the addition of an artificial foot, a prosthesis. The artificiality of articulationality, then, would be understood as coextensive with the naturality of the living organism.

That, at least, is what seems to occur in the case of the pseudopodium. For no good reason, or rather for indistinguishably good reasons—moving, eating, or both; in response to a threat from the outside, or an instinct from the inside, or both—and however provisionally, the cell alters at the same time its internal and external constitution to metamorphose from a single entity to an articulating entity: it becomes both cell and pseudopodium, whereas it was previously only a cell, and on the basis of that internal articulation, it is able to enter into an articulational relation with its environment, moving or capturing food. Subsequently, even though it reconstitutes itself, it retains henceforth that articulational relation; to paraphrase Spinoza, it will henceforth "know" something of what a body can do, and any veritable *prosthetic* relation it later enters into—grasping at food, bouncing off other bodies, attaching itself to things animate or inanimate (much later picking up tools, putting on spectacles, cloning itself)—will have been determined by that originary *articulationality*.

Freud's use of the pseudopodium analogy as a figure for the morphological transformations of libido, the temporary emission or projection of libido onto an object that changes it from an amorphous "substance" to the specific form of a relation to that object, is clearly prosthetic in its perspective: it answers to some type of supplementary desire or need, the filling of a type of deficiency. The libido, Freud tells us just before introducing the analogy in the *Introductory Lectures,* is found attached to objects and "is the expression of an effort to obtain satisfaction in connection with those objects" (*SE XVI,* 415). In "On Narcissism," just following reference to the pseudopodium, we will have been told that "the highest phase of development of which object-libido is capable is seen in the state of being in love" (*SE XIV,* 76), the symbiotic or codependent prosthesis par excellence. But object-libido is not the only case where the analogy is applicable. The question of ego-libido–object-libido cathexes relates directly to maternal–paternal versus narcissistic attachments (object choices that are either anaclitically appended, like an extended pseudopod, or remain internal, like a reintegrated one). By extension, it also relates to the ego-ideal, whereby one sets up "an ideal in himself by which he measures his actual ego" (*SE XIV,* 93), and so, in turn, to the superego, the prosthetic outgrowth turned inward and replacing the castrating father. Indeed, the metapsychological papers are in a very important sense framed on one side by the diverting of the externally directed libido toward the ego that gives rise to narcissism, and on the other by the "incomprehensible" "Economic Problem of Masochism," whereby the "instinct of destruction, which has been directed outwards, projected, can be once more introjected, turned inwards" (*SE XIX,* 164; cf. also *SE XIV,* 75–76). Freud's metapsychology is all about such catastrophic reversals.[9] From that point of view, the amoeba–pseudopodium analogy might be understood to serve as the figure for psychic operations in general, and one opines that that had to be the case once there was any theory of competing instincts whatsoever—once there was necessarily, as Freud writes in "The Economic Problem of Masochism," "a very extensive fusion and amalgamation, in varying proportions, of the two classes of instincts" (*SE XIX,* 164).

Freud's evolving topographical models of the psyche less betray, finally, a complicated spatial problem, of which descriptive versus dynamic and systematic formulations already provide explicit evidence, than elaborate the problematic I am calling that of *prosthetic articulationality*—both a matter of how something can remain indistinguishable inside and operational outside at the same time, and how something can remain organic yet function autonomously at the same time.

Yet I am also suggesting that the pseudopodial model that emerges for psychic operations in general, with its complex relational code of competing instincts and "resultant" articulations between interior and exterior and between animate and inanimate, articulations that in fact reach back into its very constitution, is also the model for life. At least it would seem to be the conception of life that Freud is wrestling with, increasingly explicitly from "On Narcissism" to *Beyond the Pleasure Principle* and beyond. And it is similarly the reason why, as we saw earlier and as we shall further develop, at certain moments in *Beyond the Pleasure Principle,* Freud continues to be in turn scientist and psychologicist, biophysicalist and evolutionist. Having first identified, as early as 1895, an intractable opposition between exogenous and endogenous stimuli that compete for priority within a supposed homogeneous organism, his scientific instincts should have led him to give priority to the overwhelming flood of external stimuli, for they appear to function according to a simple physics or logic of cause and effect. Those emanating from within, on the other hand, which are nevertheless designed in their turn to negotiate with the outside, are divided—or so it seems according to his psychoscientific intuition—between a similar logical need (hunger) and impossibly complex drives (sex) that are themselves divided between self-preservation (regeneration) and self-complication (narcissism, masochism, etc.). Indeed, the closer one examines the internal drives, the more overwhelming is the flood of stimuli that they themselves either produce or manage, and the more problematic their economy, the more difficult it is to determine what constitutes their putative efficient functioning in contrast to their putative deficient functioning. What is perhaps most complex is the series of functions that fold back in even as they extend out (narcissistic deflection of object-libido) or that produce a prosthetic appendage to replace an externally located inward-bound threat (superego). All of that, as it were, progressively bewilders Freud's sense of scientific logicity, in which he nevertheless retains a certain faith, and leads him to scale back his model to its biological origins in order to resolve the questions one way or another. The result is a tracing of the intractable contradiction itself *to*, and *in*, the origin, such that that contradiction comes to be articulated as the logic of a return to the inanimate state via the death drive. For the desired resolution to his quandary means in fact adopting contradiction as an article of faith: it means abandoning the presumed homogeneity of the organism, abandoning the exogenous–endogenous opposition—it means conceiving of life in the form of an articulation with its outside, with *some other life* but also with *some other than life*, with *pseudolife* or *nonlife*, that is always already inscribed, pseudopodium-like, inside; and it means conceiving of life as an intractable inextricability of animate and inanimate that one can try to theorize as coexisting life and death instincts, or simply lifedeath, whose origin is a nonorigin in the sense of the failure of logical reasoning to ascribe a simple originating priority of one over the other. What we call life begins as a rupture vis-à-vis itself, an interruption of inanimate by an animate that has somehow lain inert, or inanimate, within the inanimate. We cannot, from that point of view, conceive of the animate without the inanimate. Even in scientific terms, defining where animate begins—say, with the chemistry of nucleic acids—means negating, or repressing, the inanimate but still chemically active, and in a sense "animated," prehistory of life. For Freud, whether scientifically or metapsychogically inspired, it will be a matter of conceding that "the attributes of life were at some time evoked in inanimate matter by the action of a force of whose nature we can form no conception," while at the same time claiming that those *animatogenic* attributes included a force of *inanimatogenesis,* a desire to reproduce the inanimate. His "no conception" is, in fact, his tentatively projected and contradictorily conceived lifedeath drive.

If an amoeba could articulate, in the sense of speak, as it sent forth its pseudopodium, it would no doubt say "o-o-o-o"—just as Freud's Sophie's Ernst does as he throws his wooden reel, bobbin, or spool out over the edge of his cot in his "complete" game of disappearance and return. For, by the time of *Beyond the Pleasure Principle,* and Freud's grandson's experiments with the loss and retrieval of the object, with the loss and replacement of the maternal object, and hence with both the manipulation of ego-ideal and the mobilization of object-libido, the provisional or fake artificial foot of the rudimentary cell has evolved into the fully fledged automatically repetitive *fort-da* apparatus. And the pseudo-pseudopodium of the amoeba trope has become a frankly contrived and artificial semiautomatic, semiautonomous extension of Ernst's body or psyche, expressing not just how the child negotiates maternal absence and paternal interference and comes eventually to articulate its relations with the reality of a world of objects and persons but also, as the theory of the death drive develops from repetitive game to repetition compulsion, how the organic forward propulsion of the desire to live and reproduce is internally divided by an originary inorganic force willing it not to live but rather to return to mute and inert matter.

Beyond the Pleasure Principle is also clearly the metapsychological text where, particularly in chapters 4–6, Freud persists most explicitly in trying to find a biological explanation for the new facts of psychic life. He of course calls it "speculation, often far-fetched speculation" (*SE XVIII*, 24),[10] which we should by now understand as a pseudopodial proposition, hypothesis, or prosthesis that he will be ready to retract if, as he says at the very end, "it seems to be leading to no good end" (64). In those pages, however, he invites us to imagine the "living organism in its most simplified possible form as an undifferentiated vesicle of a substance that is susceptible to stimulation" (26), a sac that forms both an ectodermal crust that becomes or allows for consciousness and a protective shield to save it from an overdose of stimuli, a dead shield that has ceased "to have the structure proper to living matter [and] becomes to some degree inorganic" (27). That first introduction of the most simplified possible form of the living organism, in chapter 4, serves only as a prelude to discussion of how physical traumas and traumatic neuroses represent breaches in the protective shield, which is in turn a prelude to positing something independent of, and more primitive than, the pleasure principle, namely, the repetition compulsion. On the surface of things, that earlier discussion is not related to the treatment of the "elementary living entity" that begins immediately following exposition of the death drive in chapter 5. There, of course, it is rather a question of the organism's response to the very trauma of becoming animate: the tension that arose as a result of the attributes of life being evoked in inanimate matter "endeavoured to cancel itself out. In this way the first instinct came into being: the instinct to return to the inanimate state" (38).

And it will therefore be in order to understand the supposed subsequent emergence of self-preservative instincts that Freud will investigate the current biological theories regarding cellular development in chapter 6. But, at the very point where the death drive is formulated in chapter 5, an analogy is drawn with the emergence of consciousness that he described in chapter 4. It is immediately after being "compelled to say that '*the aim of all life is death*'" (38) that Freud makes his profession of nonknowledge, or aconceptuality concerning how life came to be: "The attributes of life were at some time evoked in inanimate matter by the action of a force of whose nature we can form no conception"; but he adds, "It may perhaps have been a process similar in type to that which later caused the development of consciousness in a particular stratum of living matter" (38).

Now, because we have been told in Freud's previous chapter that consciousness owes its existence to the development of a more or less inorganic external surface shield, indeed to a sacrifice of life—"by its death, the outer layer has saved all the deeper ones from a similar fate" (27)—the formation of consciousness is understood to come about thanks to an alternation between highly animate and inorganic matter. First, there is an undifferentiated vesicle susceptible to stimulation, an organism like an exposed nerve, the quick of pure life. It will seek a means to manage the volume of stimuli and propose two solutions: first, differentiating itself as an organ for receiving those stimuli, thereby producing the central nervous system, and second, forming a protective crust, whose "elements could undergo no further permanent modification from the passage of excitation" (26), giving rise to consciousness. That, at least, would seem to be the sequence of events Freud is describing. To survive, the unprotected organism will adopt its own reality principle and sacrifice the outer part of itself to defend its inner core, a process that we might imagine as the apotropaic gesture of an extended, and further articulated, pseudopodium, bending back around the organism like an arm hooked at the elbow to protect the face, but which then ossifies to form a prosthetic helmet. What renders consciousness possible is thus the production of a dead crust or protective shield out of highly sensitive animate matter, literally the death of that animate matter. And that is to be compared, Freud now suggests, to the inconceivable force that produces life itself out of inanimate matter. The production of inanimate from animate is comparable to the production of animate from inanimate—inconceivable indeed, except as a principle of contradictability.

On a sympathetic reading, that comparison between the origin of life and the origin of consciousness gives a type of precedence to life instincts. It suggests that life emerges from inanimate matter because there was previously another form of life in such matter: inanimate matter is not originary but is rather previously organic matter that has sacrificed itself to protect another part of itself but retained within itself the genetic trace or memory of its former organic life, which later resurrects. We might even go so far as to suggest that life survives as the unconscious of the inanimate, which might not be so very different, in turn, from our previous observation that what is called animate, biological life proceeds necessarily from an otherwise "animated" chemical activity. Freud advances that hypothesis toward the end of his text, in his gloss on Aristophanes's view of things in Plato's *Symposium,* venturing "that living substance at the time of its coming to life was torn apart into small particles, which have ever since endeavoured to reunite . . . that these instincts, in which *the chemical affinity of inanimate matter* persisted, gradually succeeded" (*SE XVIII,* 58, emphasis added). On that reading, life begins in chemistry, is sacrificed to a type of death by means of fragmentation, but resists total annihilation by external stimuli thanks to the formation of a protective cortical layer (the second sacrifice, described earlier), and as a result of transferring "the instinct for reuniting . . . to the germ cells" (58). However, just as Freud's recourse to Aristophanes's hypothesis amounts to acknowledging a scientific impasse when it comes to the death drive, we should again understand what is in play here to be a catastrophic overturning of a linear guiding logic for the emergence of life and, by extension, of the mechanist scientificity that Freud appears to want to maintain even as developments in his theory progressively undermine it. His seemingly well-meaning attempt to find the origin of the life instincts either in unicellular or multicellular organisms, pursued into the "darkness" where science cedes to mythology, will necessarily result in an impasse.

For the death drive introduces an order of existence, of life, that is by his own admission catastrophically unknown, "a force of whose nature we can form no conception," effectively holding that the first life instinct was the death instinct, that there were no attributes of life that were not always already driven by the force of death.

To return to the reading that allows contradictability, therefore, the analogical comparison between, on one hand, the inconceivable force that incited life in *inanimate matter* and, on the other, a consciousness that evolves "in a particular stratum of *living matter*" (*SE XVIII*, 38, emphasis added) enacts such a catastrophe: life that is total exposure to stimulus retreats into itself and preserves itself thanks to a partial, *inanimating* sacrifice of itself. But the death drive involves an originary inanimate that becomes animated and thereafter seeks to return to its prior state, whereas the formation of consciousness involves an animate that produces an inanimate protective crust in order to remain more controllably animate. Freud's analogical comparison—"a process similar in type" (38)—relates two processes, one of which is the reverse of the other. It amounts to a catastrophic redefinition, therefore—a *detaxonomination*—of the animate–inanimate opposition. The death drive will oblige him to abandon, correlatively, the anteriority of either life or death, the causal consecution between life and death, and the binary opposition of life and death: "if . . . we are not to abandon the hypothesis of death instincts, we must suppose them to be associated from the very first with life instincts" (57). That primary or originary "association" of death instincts and life instincts is what Derrida refers to as the successful failure of Freud's speculation, one that posits precisely a new conceptualization there where he thought he could have none:

> It [the speculative] succeeds in failing at the limit, at the moment when it is indeed a matter of going beyond the oppositional limit. Not only a given oppositional limit, but the very value of the limit as a front between two opposed terms, between two identifiable terms. For example, but these are the examples of where every term terminates, life/death. ("Speculate," 376)

The death drive is, therefore, in the origin; it is, functioning, already driving in the instinctual forces of return and repetition "which seek to conduct life into death," presumed to be "operating in protozoa *from the first*" (*SE XVIII,* 49, emphasis added). And we should understand "from the first" to mean operating *within the inconceivable force of inanimate matter that gives rise to life* as much as *at the moment of any coming to life.* What the combined sense of repetition and the inanimate gives us, from that first, is the automatism of a machine, the machine called death that Derrida analyzed in relation to the "Note on the Mystic Writing Pad."[11] That is to say, the machine called death or the machine called life, for life itself also requires the automatism of replication (for example, cellular replication) via a machine-like repetition or reproduction. The apparently strange logic that leads Freud from the compulsion to repeat to the death drive as desire to return to the inanimate state is rendered much less strange once one recognizes the inextricable thematic bond that links automatism as much to living reproductivity as to mechanical reproducibility.

The death drive is thus the inscription of a compulsion to repeat as the origin of automatism and the automatism of the origin. Life is born as the automatism of wanting to repeat the unborn experience, to repeat the inanimate experience of which it somehow has a memory. In inanimate matter, there exists a memory before memory; in the first life, there remains a memory of life before life, which precipitates a machinery of return and repetition. Life therefore comes to life as that sort of automatism, energized by an inert memory. We have to imagine that animal life proceeds from some phantom fossil that is the residue of it before it existed. We have to imagine the force of automatic repetition as a force of life functioning in inorganic matter.

Derrida calls Sigmund's Sophie's Ernst's *fort-da* game an "auto-mobile auto-affection" ("Speculate," 359). The pleasure, and desire, that takes place by means of distancing oneself by means of the bobbin, and which leads the child to repeat the experience ad infinitum, reveals the point at which the repetition or reproduction of the self engages with an other than self, as it were where the chemical *automation* of cellular matter engages with biological possibility to produce life defined—returning to the terms discussed in our previous chapter—as *automotion,* as autotelic autoaffection on the way to autodeixis. Ernst's bobbin extends his self outside of itself in order to return to itself. The child uses the prosthesis precisely to play with or explore the experience of a separation that is both his parent's separation from him and his own separation from himself. His pleasure derives from the game, and from the repetition of it, which, for Freud, ultimately means the death drive. From that point of view, the autoaffection of the game is, at one end, the organism's pleasure simply in knowing that it can move and extend itself and so be alive; in the middle, in knowing that there is a whole other world outside itself with which it can engage; and at the outside, in "knowing" beyond pleasure the drive to repeat and return to inanimate nonlife. The pleasure in it is all that all at once. The game therefore pivots on a differantial originary point of lifedeath and means that "before all else one must auto-affect oneself with one's own death (and one's self does not, before all else, exist before this movement of auto-affection), make death the auto-affection of life or life the auto-affection of death" (359). In a sense, the death drive need be no more than that simple, pseudopodial principle: for there to be life, for the self to be, it must be, or begin to be, to project in however rudimentary a form, some other; it must begin to come back round to touch, face, or caress itself from an outside itself within itself; it must be that returning to or folding back upon itself that redefines and reconfigures it, and so reenlivens it by prosthetizing it.

The death drive is that sort of primary autodeictic, and autobiographical, movement or moment of life. Freud, it seems, remained unaware of that as—with his little family scene in *Beyond the Pleasure Principle*, but also more generally—he opened the abyss of the role played by autobiography in the development of psychoanalysis. As Derrida explains, when Freud advances in that text a statement concerning psychoanalysis, even the simple fact of its existence, "he in no way is in the situation of a theoretician in the field of another science, nor is he any more in the position of an epistemologist or of a historian of the sciences" ("Speculate," 274). He is rather "attesting to an act whose producing agent and first subject the speaker knows himself to be, wishes himself to be, or alleges himself to be" (274). Much more so when it comes to his observation and analysis of Ernst's game, which cannot be understood as a theoretical argument without also being read "as an autobiography of Freud. Not simply an autobiography confiding his life to his own more or less testamentary writing, but a more or less living description of his own writing" (324). Of course, Freud is discreet in *Beyond the Pleasure Principle* about the extent of his familial involvement with the human subjects of his experiments leading to discovery of the death drive. He chanced to live under the same roof as them for some weeks (*SE XVIII*, 14). When Ernst's mother dies in a footnote, she remains the unnamed mother of an unnamed child, not Freud's daughter (16). According to the objective scientific protocols Freud is supposedly respecting, the material is no more autobiographical than if he were to refer to his observations regarding the "spinal cord of one of the lowest of the fishes," the object of research undertaken under the supervision of Brücke in the early 1880s, something he does record in his openly *Autobiographical Study* of 1925 (*SE XX,* 10). Or even, perhaps, no more autobiographical than recourse to expert or general knowledge regarding pseudopodia. But we know that the death of Sophie would haunt at least the reception of *Beyond the Pleasure Principle,* obliging Freud to insist that the book was written in 1919, while his daughter was in excellent health (all of it, that is, except the "discussion concerning the mortality or immortality of the protozoa").[12] The identity of the author of the *fort-da* game, and the less than clinical, or uncontrolled, laboratory conditions of Freud's observation of it, were always a more or less open secret. So, we can build on our comparison, developed earlier, between Ernst's *fort-da* game and Freud's repeated use of the amoeba–pseudopodia analogy—both being about the operation of object-libido—to inscribe that ana-bio-logical motif within the autobiographical nexus that Derrida has uncovered. For if Freud develops a theory of object relations on the basis—at least one of its bases—of an autobiographical familial scene which he more or less disavows, then another tributary line of inquiry, namely, the development of the object-libido, and the analogy to which he has recourse in order to explain that development, will necessarily intersect with the structure of Freudian autobiographical writing. The Freudian I or ego that observes a child playing with object relations intersects with the Freudian I or ego that plays with pseudopodia to described the emergence of the same object relations.

In the first place, therefore, the chiastic intersection of *fort-da* game and amoeba analogy confirms Derrida's insistence that a type of general-ized autobiography infects all of Freud's thinking:

Beyond . . . , therefore, is not an *example* of what is allegedly already known under the name of autobiography. . . . A "domain" is opened in which the inscription, as it is said, of a subject in his text (so many notions to be reelaborated), is also the condition for the pertinence and performance of a text, of what the text "is worth" beyond what is called an empirical subjectivity. ("Speculate," 322)

Whether such a "domain" is opened for every writing, including every scientific writing, is another question; Derrida's emphasis is on the fact of this particular "science which for once is essentially inseparable, as a science, from something like a proper name" (332). In short, autobiography in the field of knowledge that goes by the name of Freudian thinking, or in psychoanalysis, can in no way be confined to those places where Freud speaks of his own singular life experiences.

In the second place, conversely, Freud's rhetorical performances, such as his analogies, such as the pseudopodium, even where they seem not to have any autobiographical content, even where they seem not to be enounced by, or refer to, an historic "I" called Sigmund Freud, are not immune from autobiography. To repeat: when Freud says something like "consider the amoeba, it's a neat analogy of how the ego puts out some libido into an object then retracts it again," he is on one hand saying that the operation of the psyche as analyzed by psychoanalysis, the science he invented, can be compared with the objective scientific facts of zoology, but he is saying on the other hand, without saying it, that the I who has knowledge of those zoological facts by virtue of my past training and research is but a differentiated version of the same I who invented psychoanalysis, and added significantly to my theory of it by observing my grandchild at play. And he is also saying, by saying it less still, even if on another level he is saying it as explicitly as possible, that every I is an amoeba and every I is a little Ernst, including and especially the I called "Freud" who speculates his often far-fetched speculation—putting forward the hypothesis of it but consenting to retract it should you insist—in *Beyond the Pleasure Principle*.

To say that every I is an amoeba does not necessarily mean, despite the copula, that every amoeba is an I. The threshold of the class of auto-deictic autobiographical *autoi* that Derrida describes in *The Animal That*

Therefore I Am is, as suggested in chapter 1, impossible to locate, and nothing assures us that that class will include the single germ cell. Yet, as I just argued, the pseudopodium would seem, just as necessarily, to open the structure of a *pseudodactylos,* a first feint or faint attempt not only to move or capture food but also to indicate, point outside on the way to pointing back. As I also argued, the pseudopodium cannot function without the impulsion of an articulation—something that makes a part act independently of the whole—and the necessary possibility of a type of fracture: a hinge, or *brisure,* a breaking that holds together, as Derrida would have called it. But even if, in zoological terms, the autokinetic autonomy of the amoeba is to remain an open question, there is no doubt that in Freud's use of the analogy, the pseudopodium is precisely on the way to autodeixis; it is the means by which the libido self-divides and introduces relations with an object outside the organism. And that is a prelude to the formation of the superego, a form of pure deixis imported from outside so as to function as the conscience, an agent of self-directed finger-pointing par excellence.

Furthermore, if the pseudopodium is on the way to autodeixis by means of Freud's theory of object relations and the formation of the superego, it has, even more explicitly, to be an index or indicator of the way back to death. The whole theory of the death drive hangs on it, for the hypothesis itself of the death instincts will have to be abandoned unless we are to "suppose them to be associated *from the very first* with life instincts" (*SE XVIII,* 57, emphasis added). The infolded death drive, a pseudopodial hypothesis or hypothetical pseudopodium, projects from every living organism from the first—from the prokaryote as much as from the amoeba. It divides every living organism in that organism's very constitution and deforms it irrevocably.

Short of being retracted, that is. Perhaps there is no death drive, but my hypothesis, Freud says, is that there is. And if there is, it is there from the very first. So I've put it out as an idea, as a hypothesis, and I don't want to retract it. If I were to, we would be back to the beginning: there would be no death drive, just uncomplicated organisms. But, continues my liberally applied free indirect discourse, there would be no "I" either. For the reason I don't want to retract or abandon my hypothesis is that I am in it; it is mine and it is me. It is me in general terms because "the story of my life and the history of psychoanalysis . . . are intimately interwoven" (*SE XX*, 71), and it is me because the abyssally mimetic structure of my theory is such that, as Derrida analyzes, the death drive "hollows out [the pleasure principle] with a testamentary writing '*en abyme*' originarily, at the origin of the origin" ("Speculate," 304). Where life begins, or rather in what constitutes life, something about which "biology has been unable to give any certain meaning" (*SE XX*, 58), there I am with my hypothesis, about death, mine and that of life in general. My hypothesis is a hypothesis about my death, because I say *I* as I advance my hypothesis, however tentatively, ready to retract it. I say I to the extent that everyone must in uttering anything whatsoever, because in saying *I*, I indicate myself, name myself, digitize myself with a form of me that will outlive me, all of which renders me mortal; but *I* say *I* in my specific case because the scene of my hypothesis is on the far end a familial and institutional scene of inheritance, and on this end an idea about life in death before life, or death in life before life, about life beginning in reverse, arriving only to return, an idea about what, structurally and originarily, inanimates the animate. And that is why my hypothesis is also a hypothesis about the death of life in general, the death of life not in the sense of the end of life but the death in life that is born with life.

So in what constitutes life, there I am, there I am with my hypothesis, there I am with my death drive, and there I am with my autobiography. Three coextensive modes of the inanimation of life before life, projectable and retractable, projectable and retractable, projectable and retractable. Coextensive and substitutable, in fact automatically substituting one for the other at every step: a hypothesis like a tentative staking out; a death drive like a mobilization in reverse; an autobiography like a never final version-in-revision. Life gets going that way, its order catastrophically unknown.

3
THE BLUSHING MACHINE
Derrida

In chapter 1, we discussed a type of first autobiographical utterance by a Descartes who would go on to define, in very specific terms, the thinking, and, by extension, the speaking, "I." In chapter 2, we examined Freud's representation, by means of the amoebic pseudopodium, of the autokinetic lifedeath projection of the psyche that he considered to be operating in every particle of living substance. As third example of the written life of the self, we might consider what Jacques Derrida says at the very end of his 1997 conference suite on the autobiographical animal, published as *The Animal That Therefore I Am*. On its last page, Derrida, very much alive, indeed "live," extemporizing after time has run out for any more formal address, speaks of dying: "I can die, or simply leave the room" (*Animal*, 160). The immediate context of his words—presuming we knew how to interpret that term—is a reference to what Heidegger says about the capacity of *Dasein*, in contradistinction to the animal, to let be, exist, or live. The animal supposedly doesn't: "If it is the case that the animal does not comport itself toward beings as such, then behaviour involves no *letting-be* of beings as such," Heidegger writes in his *Fundamental Concepts of Metaphysics*.[1] But Derrida wants to know whether the contrary can in fact be claimed for *Dasein*. Does *Dasein* indeed let be to the extent of being radically absent from, which effectively means being dead with respect to, any "vital design" whose mobilization would impinge upon another being? Derrida: "Can one free the relation of *Dasein* (not to say "man") to beings from every living, utilitarian, perspective-making project, from every vital design, such that man himself could 'let the being be'?" And a couple of sentences later, where his example is now the sun:

> To relate to the thing such as it is in itself—supposing that it were possible—means apprehending it such as it is, such as it would be even

if I weren't there. I can die or simply leave the room; I know that it will be what it is and will remain what it is. That is why death is such an important demarcation line; it is starting from mortality and from the possibility of being dead that one can let things be such as they are, in my absence, in a way, and my presence is there only to reveal what the thing would be in my absence. So can the human do that, purely? (*Animal,* 160)

Having said that, almost at the very end of the tape of the informal session on Heidegger that closed the 1997 Cerisy conference on his work, Derrida effectively left the room, died, and let his words be. So the question to be asked is, what form of life and being do that text, and his utterance, have, and what can they tell us in more general terms about living, in particular about the autobiographical life that has been our focus up to this point? We are beholden to Derrida for shining implacable light on the testamentary status of the text, its status as remainder, *restance* or *survivance,* as discussed at length later in this chapter, something that lives on in ways that are at times common knowledge and at times totally counterintuitive; we also acknowledge his focus on the functioning of that principle of *survivance* as a type of autobiographical im*pulsion,* the desire to immortalize oneself that may or may not be explicit in a given text. In drawing on those ideas in what follows, I shall all the while be orienting things toward the different, if related, point of inquiry, discussion, or debate that Derrida most persistently returned to in later work—even though we would have to recognize it as being posed from the very beginning of his thinking—namely, the simple but insoluble question of what it is to live, of what lives.

In July 1997, in a manner of speaking, Derrida spoke, left the room, and died. Such a manner of speaking belies insoluble complexities, but certain observations are easily made. In summer 1997, Derrida was a little under six years away from being diagnosed with the pancreatic cancer that would carry him off. He was celebrating his sixty-seventh birthday, to all intents and purposes in good health, and indeed, he would survive yet another Cerisy *décade* devoted to him five years after this one.[2] However much he thought, lectured, indeed, obsessed about dying, however much it had to happen, its semiophysiological horizon was some way off. Its time remained as unanticipatable as it was necessary. As a result, the tape recording of his words from that July 1997 session remained confined to that magnetic chromium oxide archive, unexposed. Perhaps seventy-five to a hundred people heard him then, and one can imagine that only a very small handful subsequently listened to the tape. Derrida himself, we might also presume, never listened to it again. When asked for the full manuscript of his text on the animal, he systematically replied that the Heidegger section needed a great deal of work and that he didn't know when he would find the time to get back to it. We now know that that time did not remain, if by "getting back to it" we understand the formal preparation of a text for publication. Subsequently, when the question of posthumous publication of *The Animal That Therefore I Am* was raised, the status of its final chapter emerged as a not insignificant limiting factor.[3]

The first point to be made, therefore, regarding these extemporaneous words spoken by Derrida is the both naive and serious question of authority, the authority precisely of an author. The question is naive because a whole belle-lettrist tradition of textual genesis, revived most notably today in certain analyses that call themselves historicist, relies on the possibility of assigning a textual origin according to principles whose inspiration remains decidedly organicist, even creationist. According to that tradition, what lives as textual production is traceable back to its natural creative birth. The question is serious, however, because even leaving that naive tradition aside, one cannot simply ignore the distinctions between the three chapters of *The Animal That Therefore I Am* that were written and delivered in formal lecture sessions (and two of which were published during Derrida's lifetime) and the final chapter. Nor can one ignore Derrida's intention to rework that chapter and produce a book that may well have emerged in a quite different form from that of the posthumous volume now available to us. Derrida was careful to remind us that the death of the author was in many ways too reductive a concept, and that intention can never disappear from the field of differentiated utterances; simply, it can no longer be presumed to control that field.[4]

The second point, already implied by the first, concerns the operative distinction or structural difference, in the published text of *The Animal That Therefore I Am*, between portions that Derrida prepared in written form, even if the writing was designed for viva voce address, and the transcribed tape recording that is the final chapter, unwritten but made "writing" thanks to phonographic technology. That distinction is between two types of orality that are also two types of writing. Writing and voice are divided, in more ways than one. Those who were familiar with the voice of Derrida hear that voice—its tone and timbre, its humor and pathos—in different ways in each of both the formal (written) portions of the extended Cerisy lecture and the informal (transcribed) portion. Moreover, as I shall develop later in this discussion, the transcribed tape recording that would become the final chapter of *The Animal That Therefore I Am* amounts to a summary forecast of a text to come, which has its own complicated relation of oral and written life-forms, namely, Derrida's final seminar, now published as volume 2 of *The Beast and the Sovereign*. So, whether it be a matter of reading or listening, it is clear that the formal portion of *The Animal That Therefore I Am* subdivides into various levels of formality (and informality), just as the informal portion divides into various levels of informality (and formality). Both contain a whole differential terrain of discursive registers within which any simple opposition between formal and informal becomes impossible to sustain. That would be the case, in fact, with any text: there cannot be any pure discursive homogeneity, any absolutely seamless equilibrium to the utterance. If there were, one could never even raise one's voice sufficiently to utter anything at all. For those reasons, Derrida called all utterances "writing" and considered that anything that self-extends sufficiently to leave a trace consists of an uneven, ruptured, and heterogeneous mark or *trait*.

Third point: within the differential terrain of discursive registers that constitutes the final chapter of *The Animal That Therefore I Am*, the specific utterance "I can die, or simply leave the room" is no less difficult to classify, but we might hazard an attempt nevertheless. It could easily be heard functioning on the same level as various formulations from earlier in the text, such as these: "I dreamed for a long time," "I love to watch them sleep" (*Animal*, 62). It is arguably less formal, and much more homogeneous, than "the expression 'I am living (that is to say as an animal) therefore I am' is assured of no philosophical certitude" (86) and more formal than "caught naked, in silence, by the gaze of an animal, for example, the eyes of a cat, I have trouble, yes, a bad time overcoming my embarrassment" (4).

The question of the utterance's formality overlaps at a certain point with its performative status. "I can die, or simply leave the room" seems on the face of it to be uncomplicatedly constative, informing us of a matter of fact, or at least the possibility of that fact. Yet it performs in two obvious ways. In the first place, it has the rhetorical force of an example, an instantiation of "my not being there," as indicated by the context, as cited earlier: "To relate to the thing such as it is in itself . . . means apprehending it such as it is, such as it would be even if I weren't there. I can die or simply leave the room" (*Animal,* 160). One can almost hear the words "for example" being uttered between "even if I weren't there" and "I can die." In that respect, some other utterance could as well serve in its place—"I can be vaporized" or "I can go out of perceptual reach"—which means that the precise information contained in the utterance is replaceable and hence inessential, which deprives it of its constative assurance, or at least deprives it of a certain quantity of its constative content.

However, not only does the semantic field of "my not being there" expand sufficiently to allow various synonymous formulations but there also occurs a rupture in the discursive surface at the end of the preceding sentence. Not only can we hear "for example" inserted in the sentence break, signaling the opening of a paradigmatic set and the selection of two possibilities from within that set, producing something like "even if I weren't there could mean, for example, something as anodyne as leaving the room or something as absolute as death," but in "I can die," we can also hear the rhetorical effect of an apostrophe, something like this: "To relate to the thing such as it is in itself . . . means apprehending it such as it is, such as it would be even if I weren't there. *Now listen carefully and be sure to understand the full consequence of that*: I can die." It is as if Derrida were using his public philosophical discourse to convey a more or less private message, reminding whoever wanted to receive it as a type of warning that he wasn't going to be around forever.

In the second place, "I can die" is a performative utterance for reasons that will lead me to abandon this meager attempt to classify its status, its register or tone of voice (and as I have already suggested, the possibilities of hearing in it neutrality, irony, cynicism, anxiety, melancholy, and so on, constitute a whole other taxonomic cluster). Early in chapter 1 of *The Animal That Therefore I Am,* Derrida recounts how it is, and how *one is,* indeed, how *therefore I am,* when I find myself naked before the eyes of my cat in the morning. In accordance with his gesture of refusing the massive totalization of millions of animal species that inheres in the opposition human–animal, Derrida wants the cat that looks at him in his animal nakedness and human shame to be recognized as "*this* irreplaceable living being. . . . What we have here is an existence that refuses to be conceptualized." However, he cannot at the same time avoid infecting the cat with mortality: "[it is] a mortal existence, for from the moment that it has a name, its name survives it. It signs its potential disappearance. Mine also, and that disappearance, from this moment to that . . . is announced each time that . . . one of us leaves the room" (*Animal*, 9).[5] In other words, not only does the subject that says "I can die" affirm its mortality but so does the cat once it is recognized as a singular existence, for example, by being given a name; he/she/it receives that name and receives the status of a singular irreplaceable existence as the announcement of his/her/its own death. Mortality is performed each time one of us leaves the room, but it is also performed by every means by which the singular existence of each of us animals is affirmed, which effectively means with each and every member of what we might call the very open set of our personal utterances. It is especially performed, in any case, when one of us says "I." Before adding "can die or simply leave the room," "I" have already performed, simply by uttering the first person pronoun as a signifier of my singular irreplaceable existence, the necessary possibility of my death; I have already left life and the room to the extent of opening the structure of my definitive departure. *Leaving or dying is no longer something to come solely in an unpredictable future, but something that already infects the very present of my utterance.* So the "I can die, or simply leave the room" of the final chapter of *The Animal That Therefore I Am* comes to be riven by the abyssal effect of a performance of mortality and the announcement of death: *I, who not only can die, but who is dying at least a little by saying "I," can die, or simply leave the room, which is another way of dying.*

Dying, indeed, is explicitly at stake in the next paragraph of that first chapter, where Derrida continues, reinforcing the sense of shame in terms to which I'll return: "But I must immediately emphasize the fact that this shame that is ashamed of itself is more intense when I am not alone with the pussycat in the room. Then I am no longer sure before whom I am so numbed with shame. . . . In such moments, on the edge of the thing, in the imminence of the best or the worst, when anything can happen, when *I can die* of shame or pleasure, I no longer know in whose or in what direction to throw myself" (*Animal*, 9–10, emphasis added). Thus, beyond all the questions just raised regarding the status of the "I can die" of the final page of the book—the instability of discursive register, the performance of an example, and the abyss of mortality and death—there also exists this uncanny repetition, in the opening pages of Derrida's text, of what is stated on its final page: *one of us can leave the room and I can die* ("[its] disappearance is announced each time that one of us leaves the room. . . . In such moments . . . I can die of shame or pleasure"). Now that might simply suggest something as banal as a rhetorical tic whereby leaving the room is Derrida's everyday lifetime figure for dying (his superstitions regarding leaving the house, for example, are well known[6]). However, it necessarily reminds us of what functions as the basis of the whole performative apparatus I have been describing, namely, the iterability of the utterance. However different the two contexts in which the repetition of "I can die" occurs, however semantically incommensurable the two versions, however accidental, however contingent upon the contrast between prepared written text and transcribed oral extemporization that I have just discussed, they do in fact appear in the common context of a single textual format that is the book entitled *The Animal That Therefore I Am*. And it is that disseminative citationality, the utterance's potential for being cited in a radically different and unsaturable context about which Derrida was so categorical in "Signature Event Context," that finally ruins the possibility of ascribing to the sentence "I can die, or simply leave the room" anything like taxonomic exactitude.

Granted, a repetition is not the same as iterability, which is the structure of repeatability that invades even the supposed single utterance. Similarly, it is not the chance of this fragmented repetition alone that loosens the utterance "I can die" from its moorings in the ways I have described. Still, the fact that "I can die" is uttered twice, each time in its own context, but together in the broader context of the whole book, necessarily creates an echo whereby each instance is heard in the other; and it brings the perhaps casual remark of Derrida's improvised sketch of the final chapter back to the terms being developed in and from the beginning of his address, where the operative theme is not, as at the end, the letting-be of which *Dasein,* unlike the animal, is supposedly capable but rather the sense of nakedness and concomitant shame specific to the human. *In my more intense shame,* Derrida writes, *on the edge of the thing, I can die of shame or pleasure.*

Besides, the repetition of the syntagm "I can die" enacts a more specific function of iterability that interests me here, namely, its technology or automaticity. Once an utterance is severed from its producer to the extent of being repeated in quite diverse contexts, such as is the case here, capable of being isolated for the simple effect of the repetition itself by being extracted from the larger syntagmatic chain or flow that surrounds and neutralizes it, then language begins to sound or look like a machine at work. It suddenly has inscribed within it a lifeless automatism, finds itself reduced to the smaller or larger syntagmatic elements that we know it to be constituted of—at base a small set of phonemes whose permutations are repeated ad infinitum—combining here to form the clause "I can die" like a pleonastic mantra no longer spoken by a Derrida referring either to how to really let things be or how intense his shame is but instead intoned by linguistic technology itself. By intoning "I can die," the linguistic machine would perhaps be repeating something Derrida wrote a long time ago, in "Freud and the Scene of Writing": "The machine is dead. It is death. Not because we risk death in playing with machines, but because the origin of machines is the relation to death";[7] or perhaps, conversely, the linguistic machine would be declaring its mortality or lethality precisely to give itself a type of life, the type of life we might imagine to animate the machines we call biotechnologies—in one case, then, a lifeless origin, in the other, no clear indication of where or when it will grind to a halt.

"I can die" echoes across the far from empty space of opening and closing pages of *The Animal That Therefore I Am.* It echoes, and as it were *lives,* something of an animal of an utterance, or at least an animate utterance, even if it be technologically animate. If I am paying so much attention to the utterance, it is, in the first place, because the principal ethical exhortation of Derrida's book is that we attend to the singularity and irreplaceability of *whatever* lives, *however* it lives. Following that, the sentence also echoes through the question that Derrida raises and critiques more than once, concerning philosophy's habit of dividing between the human and every other animal species by means of any number of criteria, certain of which are presumed to be logically related: the relation to death, precisely, and thus to burial and mourning, but also the relation to language and how that functions as a difference between reaction and response. The repeated and recontextualized "I can die" also poses the question of reaction versus response. In this repetition, are we to understand that one version or context of the utterance is reacting or responding to the other? This is not the same question as whether Derrida is reacting or responding to himself; rather, it involves asking further what it means to presume that the automatism of the linguistic machine derives from a lifeless repetition, which can be rigorously separated from an animal reaction, which can in turn be rigorously separated from a human response. I seemed to begin to answer that question by suggesting that there is more to the repetition than a simple repetition, more than a "parroting" of one version by the other, and that what exceeds repetition does not reduce to the manipulation of a canny reader extracting three words from one context and relating them to another context, any more than it reduces to the conscious or unconscious intent of an author prompting such a reader. The "I can die" of page 160 is not, to use Lacan's term, simply a "relay" of what was said on page 10.[8] Or, as Derrida's adestinational postal principle insists, there can be no such thing as a simple relay (cf. *Post Card*, 66–67). The very principle of linguistic *iteration*, all by itself and before we come to accept a concept or principle of *iterability*, would be posited on the basis of a type of response. Every time we read or hear a repetition in language, beginning with an alliteration, assonance, or rhyme, and going all the way to rhetorical emphases and thematic motifs, we receive them as the text's responding to itself and so animating or livening itself, calling and responding to itself as though it were conversing with, singing, or orating to itself. What iterability adds to that idea, transforming it in the process, is the insistence that there is harbored within such repeatability the irreducible automatism that I have just described, rewriting language's self-re*sponse* as an auto*sponta*neity—language functioning *sponte sua*, of its own accord—which is a mode of the automotricity or autokinesis that we understand to be at work in every life-form. Iterability means that language moves itself beyond itself at its very origin; that such movement or autodisplacement, such a rupture within the intact closed circuit of nonmeaning, is what produces sense and gives to language its force of signification.

Now that is not to say that language is the same form of life as a protozoon, on one hand, or a human being, on the other, or that it lives in the same way as any of the millions of life-forms that exist between the two. It is to insist, again, that criteria of distinction such as react-ability or respons-*ability* founder as means to divide animal from human once the repeatability of iterability is discovered disturbing the limits of one and the other. But that insistence on the fallibility of the reaction–response distinction as a means of separating human life from animal life also implies a disturbance of limits between life and so-called technological nonlife. Autokinetic iterability, and the "duplicity" it entails, is the basis of Derrida's insistence, in 1993, on there being technics from the beginning, at the origin and especially in the origin of what we call the human:

> Any living being, in fact, undoes the opposition between *physis* and *technè*. As a self-relation, as activity and reactivity, as differential force, and repetition, life is always already inhabited by technicization . . . a prosthetic strategy of repetition inhabits the very moment of life: life is a process of self-replacement, the handing-down of life is a *mechanike*, a form of technics. Not only then, is technics not in opposition to life, it also haunts it from the very beginning.[9]

Much of Derrida's writing, especially in the last ten years of his life, was explicitly involved in thinking that fact of life. It is the arc of a trajectory that connects the animal of 1997 to the wheel of *Rogues* (2002) and *The Beast and the Sovereign* seminars (2001–3), and the learning to live, finally, of the *Le Monde* interview from just before his death.[10] But, as he was wont to insist, there had never been any other question for him. I mentioned in my introduction that in *For What Tomorrow*, a series of interviews with Elisabeth Roudinesco from 2001, he begins discussing the "Politics of Difference" by asking for a step to be taken back, reminding us that diff*e*rance was never about anything more than a rethinking of life: "What is universalizable about *différance* with regard to differences is that it allows one to think the process of differentiation beyond every kind of limit: whether it is a matter of cultural, national, linguistic, or even human limits. There is *différance* as soon as there is a living trace, a relation of life/death or presence/absence."[11] There is *différance* as soon as there is a living trace, but as soon as there is *différance* or the trace, then the question of what lives, how it lives, or what life is is irrevocably posed and interminably problematized. The trace of or as *différance* is a strange and complicated life-form—if we can call it that—both a remainder, as with a footprint, of something that self-extends in autokinetic spontaneity and an inanimate inscription such as could be stamped by a machine; both the chance mark whose producer is dead, lost, unidentifiable, and irretrievable and the minimal impulse that gives rise at the origin to the origin of all things. Without it, there could be nothing like what we call life; because of it, what we call life has from the beginning left something we would normally call dead behind it, without for all that divorcing itself, in the sense of an absent–present opposition, from what it has left behind.

One would be hard-pressed to decide, on that basis, which of the two versions of "I can die" has the most life in it: the earlier, formal, more "accidental" version, which, on the surface, seems to have less of an authorial investment, less rhetorical force, and instead appears to have detachment inscribed within it, which would consequently define it as already living free of its source; or the later, vernacular, exemplary version, which is replete with familiar overtones—recognizable to a greater or lesser extent, depending on the level of one's familiarity with Derrida's speaking and writing—and so yawns like an emotive chasm in the wake of his disappearance, as if speaking its own life as no less precarious than tenacious? In each case, the life of the utterance would be derived precisely from the radical letting-be of a type of death, a letting-be that is a letting-function-on-its-own, under its own steam, as we say, in a type of technological automaticity.

Before or beyond anything else, therefore, that type of automatic self-generation of "I can die" operates across the pages of Derrida's text as the very structure, or force perhaps, that allows for the other forms of echo that one can hear in it. We can therefore recast what we were saying at the beginning about hearing the tone of Derrida's voice. Before hearing any reminiscent voice of, for example, a departed friend or colleague, one hears, by definition, the death that attaches itself to every utterance the moment it leaves the mouth of its producer. The voice of Derrida that can be heard, with one tone in the formal lecture of chapter 1 of *The Animal That Therefore I Am* and with a different tone in the informal presentation of chapter 4, is necessarily overlaid with an affect or pathos that derives from its becoming the living voice of a dead person. And those effects make the text live on within a particular context of academic exchange and human friendship. But that form of living on, however emotive it be, is a function of the general structure of automatic iterability and of the machine of death at work in every sign whatsoever.

Hence this *retroversal* echo, in the early pages of *The Animal That Therefore I Am,* of the seemingly offhand formulation of its last page, brings us thick into the nexus of animal life, death, and putatively inanimate technology. What mobilizes that nexus, at least in the context of the first "I can die," what serves as its motor, is shame, and what produces shame, its generator, is nakedness. In the autobiographical animal lectures, nakedness is Derrida's subject from the very beginning. First words: "In the beginning, I would like to entrust myself to words that, were it possible, would be naked. Naked in the first place—but this is in order to announce already that I plan to speak endlessly of nudity" (*Animal,* 1). Nakedness is the state of the beginning, the original animal state of the human, subsequent to which—as we know, the precise order of events is unclear—the human animal will "fall" into its superior status as lord over all others. Knowledge of or shame regarding nakedness is the basis for that promotion of the human to a clothed, technological animal. In that sense, it might be understood as the first anthropoautobiographical moment, the point at which a particular animal inscribes itself as human. We saw it operating in chapter 1, where a fully conscious and clothed Descartes (even though, as we shall later see, he often appears not to have dressed warmly enough) sought to mask his shame as he readied himself to embark upon the stage of the world and to set in train his conceptual apparatus, installing in the same movement a type of generalized masking or originary prostheticity that would rob the *larvatus prodeo* of its autocritical awareness as much as it would, by extension or protension, rob the *cogito* of its organic foundation.

Shame is the complicated system of self-reflection that begins with consciousness of our nakedness. No animal knows it is naked. As Derrida says, it "is not naked because it is naked. It doesn't feel its own nudity.

There is no nudity 'in nature'" (*Animal*, 5). The human knowledge of nakedness is necessarily produced, logically, and "historically" or mythologically, by means of an encounter with nonnaked, nonclothed animality. And a strange (non-)encounter is what it necessarily will be. No doubt it doesn't occur when one happens upon every single protozoon, nor even necessarily when we encounter a snake in an Edenic garden; but, as Derrida explains in the early paragraphs of *The Animal That Therefore I Am*, where leaving the room and dying come up, we are ashamed in front of an animal, such as a cat, who, by looking at us, tells us that he or she knows that we know that naked we all came into the world and naked we shall return. It is on the basis of such knowledge that we take animals as our companions; name, feed, and in some cases clothe, shelter, and eventually bury them; or, in the case of other animals, industrially farm them, bioengineer them, kill them more or less humanely, and eventually eat them.

Now, for a nonnaked, nonclothed cat to tell us, by looking at us, that he or she knows that we know that we are naked, the cat would in fact have to be something of a snake. A real cat, not having any idea about nakedness at all, not having such a concept, couldn't possibly tell us that. Nor, as the Genesis myth suggests, is consciousness of nakedness something that can arise spontaneously, as it were, out of nature. Someone has to bring us to that point—someone, some animal, or some thing. Something *super*-natural. Imagine, after all, some prelapsarian and unadorned human animal taking a stroll through a pristine garden and deciding, on the basis of the good advice she seems to be getting, to think nakedness. She would have first to invent the concept out of whole cloth, not having any dialectical foundation for it, not knowing what nakedness was any more than nonnakedness. This means inventing the concept tout court, to begin with, inventing the possibility of a dialectical opposition such as that between nakedness and nonnakedness where before there was only differential hirsuteness or pilosity, degrees of hair or fur. Presuming she got that far, however, she would have to clothe her nonnakedness in the concept of nakedness, to place nakedness like a covering over her originary nudity, which would require her at the same time to invent the concept of covering or clothing as a derivative of the concept of nakedness, which is itself, as I have just argued, a qualitative derivative leap from the state of the nonconcept. No mean feat therefore. Indeed, a universal overturning is required to set in motion the dialectical conceptual apparatus itself, to introduce what we understand as knowledge. That sort of heavy industry or high technology is what we call the fall. Before we "fall" into consciousness of nakedness, of good and evil, before we fall into shame and sin, nature has to have already fallen out of itself into nonnature, into a technology of conceptualization. From that perspective, shame comes to be seen as the conceptual machinery itself, a machine set in motion by itself, always already on. To our shame and to our credit—presuming we could hear such terms as neutrally as possible—the human "reacts" to knowledge of animal nakedness by developing for itself an infinitely expandable prosthetic technology, from clothes to cover that nakedness and shame all the way to the techniques of domestication and domination, agricultural industrialization, genetic and other experimentation that I have just evoked, which finally risks adding up, in Derrida's estimation, to something comparable to a genocidal technology of death to the animal (*Animal*, 25–26). But, to repeat what I am arguing here, those technologies do not in fact begin with the artisanship of attire, for a conceptual and dialectical technology of enrobing, like a masking, will have been in operation from first blush, in that very blush. Nonnakedness as concealing of nakedness will have had to be a structure of nonnakedness as lack of consciousness of nakedness; there will have had to be the technology of that perverse or impossible dialectic within pure nudity itself.

Of course, Derrida will also argue that shame and technology are finally no more pure dividing lines to distinguish the human on one side from the animal on the other than are any of the other criteria to which philosophy has consistently had recourse to impose its reductive delineations. Can we rigorously determine, he asks, that the animal is deprived of language, clothing, laughter, mourning, boredom, deceit, music, hospitality, the gift, and so on? But more precisely, "if one takes into account . . . a seduction that is tenderly or violently appropriative, one can no longer dissociate the moment of sexual parade from an exhibition, or exhibition from a simulation, or simulation from a dissimulation, or the dissimulative ruse from some experience of nakedness, or nakedness from some type of modesty . . . or shame" (*Animal*, 60). Thus it would be not only the human animal, but also many other animals, and potentially every sexed animal whose automotricity includes mating, that thereby participated in some sort of mechanics of shame and so defined itself as originarily technological. Perhaps as soon as an animal no longer reproduces simply by, as it were, copulating with whatever it bumps into, but instead allows itself to be seduced, to veer off track toward a mate, and even more obviously once it involves itself in any sort of mating game whatsoever, then that animal is clothing its habits in a type of technology; that mating becomes a technology, however natural, however much a programmed animal reaction we still might consider it to be. And no doubt the same could be said for any ruse whatsoever by which an animal does other essential things, such as obtain its food, however instinctual we might consider that to be.

As I intimated earlier, the "I can die" of the end of *The Animal That Therefore I Am,* and the Heideggerian analysis of letting be as such, receive extensive elaboration in *The Beast and the Sovereign II.* That seminar, and Derrida's formal teaching career, closes in March 2003—one week after the American invasion of Iraq and seven weeks before Derrida's cancer diagnosis—with the question "who can die?": "The question, that was the question of the seminar, remains entire: namely that of knowing who can die" (*B&S II,* 290). With that, he left the room and began to die. Thus, in one sense, the iterated "I" who can die near the beginning and at the end of *The Animal That Therefore I Am* returns at the end of *The Beast and the Sovereign II.* But it returns with somewhat less assurance, and the "death itself, if there be any, [that] was our theme" (290) throughout the seminar reverberates as a question regarding each term of that simple syntagmatic chain: regarding capability, power, even sovereignty (who is capable of death, who can manage death [*qui peut la mort*]?); regarding the subject (who/I/the animal); and regarding the immense, and immensely undecidable, question of whatever lives, of what is called "the living [*le vivant*]" that was announced in the first volume.[12] But in a stronger sense, that "I" receives a further, affirmative iteration through the fantastic and phantasmatic possibility that *I can die alive* or *live dead,* which derives from the structure of survival or what Derrida prefers to call *survivance*: "a survivance or a surviving (but I prefer the middle voice 'survivance' to the active voice of the active infinitive 'to survive' or the substantializing substantive survival)" (*B&S II,* 131).

Derrida's focus in the second year of the seminar is on two texts. The first is Heidegger's *Fundamental Concepts of Metaphysics: World, Finitude, Solitude,* with its distinction among worldless *(weltlos)* stone, impoverished *(weltarm)* animal, and building *(weltbildend)* human and between a *Dasein* that dies *(sterben)* and an animal that simply comes to an end *(verenden).* In that context, for Heidegger, whereas both human and animal can simply leave the room, only the human can die.[13] But, as Derrida points out, being capable of dying, having access to one's death as such is by no means determined or assured by having the word "death" in one's lexicon, by having the habit of seeing others die, or by possessing the "near-certainty that . . . *we can die* from one moment to the next" (*B&S II,* 117, emphasis added, translation modified).

Derrida's second text is *Robinson Crusoe,* in the context of which, particularly in sessions 3–5, he effectively rephrases the hypothesis "I can die"—via Crusoe's fear or phantasm—as "I can be buried alive." Whereas that leads in the first place to an extensive analysis of the two modes of disposal of dead remains that are available for choice in Western democracies, namely, inhumation and incineration, it gives rise in the second place to sustained reflection on the sense of what, for my purposes, constitutes autobiographical survival. If I can die, in what manner or form can I survive? What form of life is survival? What is the life of what survives me? What is the form of my life in what, and indeed in whom, survives me? When I have died, what is the living status of my remains, of what remains? For if I can never, as it were, empirically say "I have died," there is a sense in which I can, often do, even necessarily say "I am buried alive."

I speak from the perspective or phantasm of being buried alive when I leave instructions for the music at my funeral, or concerning how my remains are to be disposed of, for one can "only speculate on his or her own death on the basis of the imagination or the phantasm of the living dead, or, at the limit, of the dead one still alive enough *to see him or herself* die and be buried" (*B&S II,* 160). But I also speak, act, perform from the perspective of being buried alive in more radical terms to the extent that whatever general organization of "life" I undertake with a view to my departure—from explaining how to decalcify the espresso machine or putting order in my finances to bringing a child into the world, and everything in between—is a means of being present at my death; it means both acting, playing, or being dead while I am still alive and being alive after I am dead, participating in what Derrida here calls the "spectrality of the living dead" (117). Indeed, in the most general sense, "our thoughts of death are always, structurally, thoughts of survival. . . . The logic of this banality of survival that begins even before our death is that of a survival of the remainder, the remains, that *does not even wait for death to make death and life indissociable*" (117, emphasis added). At the outside, therefore, the human concept of death, or the ontological concept of death, is necessarily a concept of survival. That would lead us to conclude, on one hand, that even the most atheistic thinking of death intersects at a certain point with religious thinking concerning death as afterlife and, on the other hand, that the prospect of death that invades our daily existence—from dread of fatal accident or terminal illness or unfathomable nothingness to fear of falling, drowning, or being buried alive, but also and especially by means of every artifact and every archive, organic or inorganic, every tree we plant and every dream or memory we are responsible for, everything we make or write, everything we produce in life to outlast us—functions as a type of inanimation, the structural inclusion or incursion, within life, of inanimate remains. It is only by dealing with those surviving remains that we are able to live.

Beyond all that, however, I am buried alive by the generality that Derrida calls the "almightiness [*toute-puissance*]" of the fantasmatic virtuality of dying alive, "an almightiness that . . . rules over everything we call life and death, life death" (130). That is to say that even if dying alive were possible only as fear or phantasm, such as is Robinson Crusoe's "great organizing fantasy (terror and desire)" (117) while he is alone on his island, even if, as is the case, one cannot, in fact and in the present, be both dead and alive, that possibility, once admitted, still functions as a type of power over life and death, governing life and death to the extent of reorganizing and blurring the distinction between them. Even if that possibility were nothing but a fiction, it would still be the fact of a fiction that never stopped haunting fact and the power of whose haunting would be impossible to control. For that is indeed what Derrida finds put on stage in the fiction *Robinson Crusoe*.

But if the character Robinson Crusoe performs the autobiographical narrative of an "I" who can leave the room or die by virtue of his fear or fantasy of being buried alive, *Robinson Crusoe* as work of fiction is a form of living remains by means of which the author Daniel Defoe survives dead. That is the obvious fact of signatory or testamentary writing already referred to, but it is also a type of paradigmatic instance of the generalizable principle of *survivance* as lifedeath being developed here:

> Now this survival, thanks to which the book bearing this title has come down to us, has been read and will be read, interpreted, taught, saved, translated, reprinted, illustrated, filmed, kept alive by millions of inheritors—this survival is indeed that of the living dead. As is indeed any trace, in the sense I give this word and concept, a book is living dead, buried alive and swallowed up alive. . . . The book lives its beautiful death. That's also finitude, the chance and the threat of finitude, this alliance of the dead and the living. I shall say that this finitude is *survivance*. Survivance in a sense of survival that is neither life nor death pure and simple, a sense that is not thinkable on the basis of the opposition between life and death. . . . This survivance is broached from the moment of the first trace that is supposed to engender the writing of a book. From the first breath, this archive as survivance is at work. But once again, this is the case not only for books, for writing, or for the archive in the current sense, but for everything from which the tissue of living experience is woven, through and through. A weave of survival, like death in life or life in death. (130–32)

According to the logic already developed, that tissue of living experience would begin with any minimal utterance by which an irreplaceable existence has inscribed upon it or recognizes its own mortality, for example, in being named or in saying "I." Autobiography, whether accidental or institutionally formatted, whether as remnant or as ordered narrative, would be one name for a certain material or visible weave of that tissue, its becoming explicit, its archival deposit. But my argument here is that autobiographical tracing is a function or structure whose minimality is impossible to determine and that reaches back to wherever and however life first speaks as life. Then already we are dealing with an "I" who can leave the room or die; then already I live in the form or space of one who leaves remains. One who(se) remains therefore (are) living dead.

Beyond the specific graphic instance of a living *autos,* or of a self-formed life that produces something called a fiction or a book, beyond the historicoliterary institution of autobiography (whether or not that be generically distinguishable from many other forms: confession, memoir, journal, testimony, testament), the *survivance* effect that Derrida describes is articulated specifically through the fact of a living body's becoming a dead thing. At that point I say to those who survive me, in the contemporary Western context, one of two things: either bury me, after first making sure that I am well and truly dead, and then deal with the fact that I'll always stay present in my grave; or burn me, take away my fear of being buried alive or of rotting slowly, annihilate me (with anguish or pleasure) and deal with the fact that for being nowhere specific I am left everywhere about you. Those two regimes of disposal represent two very different approaches to the dead body as inanimate object, but each produces its own form of spectral survival. But what goes for the living body become dead thing itself goes also for every other thing that survives the living life of that body, every thing that functioned prosthetic to it throughout its lifetime but has not now died along with it: its clothes, its belongings, its books, its papers, the whole archive of it referred to earlier, but that extends, as I noted there, along lines that do not clearly divide between inanimate and animate, between inorganic and organic, including, for example, plants and students, hopes and memories. All that also—or what there is of the dead one that still lives in it—calls to stay alive or to be buried or burned. For in the final analysis, what survives as more spectral, more indissociably dead and alive than the dead one itself, undecidably produced by the one who dies or those who remain, and already at work well before any actual death occurs, is the form of life called mourning: "mourning does not wait for death, it is the very essence of the experience of the other as other, of the inaccessible alterity that one can only lose in loving it—or just as much in hating it" (*B&S II,* 168).

The production of remains is thus without a *situable* origin and without a *saturable* range. In that it obeys the law of any utterable trace that Derrida describes in "Signature Event Context," its field of operation being in no way limited to what might be foreseen within the event of death in the usual sense.[14] Remains survive at the discretion of the survivors—they may be neglected or ignored, preserved or destroyed—but the survivors themselves constitute an open set. And in both a strict and a general sense, remains, like the elements of a linguistic utterance, are iterable. They may be repeated in the strict sense to the extent that they constitute a form of artifact, such as a book or even ashes, which makes them susceptible to artisanal or mechanical reproduction, as well as misuse or forgery, and they may be repeated in the general sense inasmuch as they are less produced than "fall" within the structure of whatever is uttered, obtaining their significance from an iterability that makes them always already the trace—something like the symmetrical yet disjunctive reverse side—of themselves.

Because, as I explained earlier, iterability carries with it a form of automatism, Derrida will refer explicitly to the *survivance* effect, specifically in the passage on the book that I just quoted from at length, as a machine, in fact as the originary *technē* that disturbs the very integrity of the living: "And the machination of this machine, the origin of all *technē* . . . is that each time we trace a trace, each time a trace, however singular, is left behind, and even before we trace it actively or deliberately, a gestural, verbal, written, or other trace, well, this machinality virtually entrusts the trace to the sur-vival in which the opposition of the living and the dead loses and must lose all pertinence, all its edge" (*B&S II*, 130). What survives is no longer an organic part of the living being that it derives from; it is irrevocably separated from that form of the living, "cut off" from it, as Derrida explains in a different context. Of course, it will thereafter have a complicated prosthetic existence vis-à-vis what it derives from, and knowing what to call that, asking to what extent it can be called a life, is the point of this discussion and this book. Paradoxically, however, it is precisely the fact of being cut off that gives to what survives that "sort of archival independence or autonomy that is quasi-machinelike . . . a power of repetition, repeatability, iterability, serial and prosthetic substitution of self for self."[15] The apparent enfranchisement, into a life of its own, of what lives beyond what lives, constitutes the quasi-machinelike iterability of something that will go on functioning somewhat automatically as long as it can be *invoked* or *engaged* in any way whatsoever, with or without reference to its original source. Remains may always be the remains of something, but they keep on remaining, like some artificial life-form, irrespective of whatever they are the remains of. And that structure of remaining reaches back to "infect" or "mortify" the living being or thing that will be *remaindered,* unable ever to be excluded from the life from which the remains will come to be derived. Such a life or living being will therefore always harbor within it not just its death-to-come but also its quasi-machinelike iterability or survival to come. In that respect, the iterable trace is not just something left behind after life, in the sense of being deposited outside life, but also a strain that runs through it—not just the dead remainder of what lives but also the automatism within it. That is what makes it living dead, lifedeath, a fraying of the edge between living and nonliving.

Obviously there is no clear mechanical model for such a machine, or quasi-machine, for originary *technē* itself, but in later work, beginning particularly with *Rogues,* Derrida increasingly explains the "decisive power . . . [of] this logic of iterability" (75) by means of recourse to the wheel. The wheel moves outside of itself while turning upon itself; it distances while returning to itself. Although the wheel is never explicitly called the machine of life, it functions on one hand as "a sort of incorporated figural possibility . . . for all bodily movements as physical movements of return to self, auto-deictics, autonomous but physical and corporeal movements of auto-reference" (75), and, on the other hand as the machine of *ipseity* in general, of the working of the self or *autos* that Derrida concentrates on as the basis of what we call sovereignty. Robinson Crusoe wrestles with reinventing the wheel at the same time as he wrestles with reinventing himself, for example, a self that is at the same time sovereign and subject; but he is also learning, as if for the first time, to deal with "an autonomization, and automatization in which the pure spontaneity of movement can no longer be differentiated from a mechanization, a progress in the mechanization of an apparatus that moves by itself, auto-matically, on its own, toward itself at the moment it travels toward the other. . . . This mechanizing and automatizing autonomization . . . is not without relation, at least an analogical relation, with what is called sovereignty" (78). Crusoe's autobiographical account of being marooned is, in the first instance, the account that we read in the book of the same name. But beyond that, it is a story of his *rehominization,* of the reliving of the phylogenetic development of the human species that takes place through not only his relations to other animals and eventually to another human but also through his body's relation to itself and to technical apparatuses such as his wheelbarrow. And beyond that, as well as by virtue of that, Crusoe's autobiography is the story of the "development" of the self, which is staged by every single autodeictic tracing whose minimal instance is impossible to determine because it is at work on the very *livingdead* edge of life.

In every instance of such tracing, as in every version or "level" of Crusoe's autobiography, the man, the human, the self is involved in a relation to the machine: the machine of his narrative, the wheelbarrow, and the automatizing autonomization of the *autos*. As Crusoe rolls out each new episode of his narrative, as he rolls out his brand-new cask or grindstone, and as he rolls out his pact of sovereignty to be sealed with the wild beasts, or with Poll, or with Friday, he is both extending by supplementary additions the scope of his influence—his power to be, his power to live—and dividing the same by entrusting it to forms of inanimation: to prostheses that work not just as external appendages but as the artificial or automatic enlivening of the self that starts from the very moment it ventures out of itself and into the world:

> I am thinking . . . of a structural configuration . . . in which everything that can happen to the *autos* is indissociable from what happens *in the world* through the prosthetization of an ipseity which at once divides that ipseity, dislocates it, and inscribes itself outside itself in the world. . . . The wheel is not only a technical machine, it is in the world, it is outside the conscious interiority of the *ipse,* and what I want to say is that there is no ipseity without this prostheticity in the world. (*B&S II,* 88)

Crusoe's automatic prosthetization, by means of narrative, machine, self-exteriorization, is of course a form of protection, either personal or vicarious: we are to learn from his experience of self-exposure and vulnerability; he will build a new world in which to survive; there will be developed a new personal and political order. A prosthesis replaces what is lacking or defective in order to restore a presumed original integrity of the organ or organism (an integrity that—it is my unyielding argument here—never in fact existed). And, as we are now required to understand, prosthetic intervention is increasingly inseparable from genetic intervention, the machine is increasingly a nano- or biomachine. Hence, if Derrida employs the figure of the wheel to insist on the rudimentary mechanical and material sense of the originary transmutation of the human—its leap into automobility—he also bioengineers that figure to have it stand for a different version of the same livingdead tracing that has been prominent in work of the last ten years of his life, namely, autoimmunity. The figure of the wheel is "the turn of a trope that constructs and instructs in the relation to self, in the auto-nomy of ipseity, the possibility for unheard-of chances and threats, of automobility, but also, by the same token, of that threatening auto-affection that is called autoimmunity in general" (*B&S II*, 75). Failing the wheel, therefore, we can find in autoimmunity a more organic form of the machine of life, the automatism by which life prosthetizes itself in order to extend itself, protecting and threatening itself in the same movement: "autoimmunity consisting for a living body in itself destroying, in enigmatic fashion, its own immunitary defenses, in auto-affecting itself, then, in an irrepressibly mechanical and apparently spontaneous, auto-matic fashion" (83). Autoimmunity may be understood in that sense as an organic machine, as the organism's self-regulating bioengineering, but one that is in operation from the first turn of the wheel of organic life—from life's first rollout of its prototype. Life's first autoimmune response, we might say, is the reaction to itself as machine, to the fear that it is being invaded from within, for example, by the very automatism of its necessary autogeneration.

In the context of these remarks in *The Beast and the Sovereign II*, Derrida includes a whole other question in the structural configuration of autonomic automatization that he is developing. That question is prayer, which Crusoe sets in parallel with the wheel: he needs to relearn how to pray in the same way that he needs to reinvent the wheel. Though the relation might seem at first sight incongruous, Derrida finds prayer to be represented in *Robinson Crusoe* as "a cry that is almost automatic, ir-repressible, machinelike, mechanical. . . . In both cases [of prayer and the wheel] we would be dealing with an autonomization, an automatization in which the pure spontaneity of movement can no longer be distinguished from a mechanization" (*B&S II*, 77–78). For prayer points to the general-ized autoimmunity whose mechanism Michael Naas has analyzed in such lucid detail in a context where once again the machine appears to infect the question of religion. In his brilliantly sustained reading of Derrida's

"Faith and Knowledge," Naas shows how the emphasis Derrida places there on technology does not just underline a problematic of the contemporary religious scene but points to a broader phenomenon: how whatever seeks to protect itself as unscathed finds itself dealing with effects of the machine and practicing a form of "auto-immune auto-indemnification."[16] Thus, when Derrida writes, in that context, that "the reaction to the machine is as automatic (and thus machinal) as life itself,"[17] he means not only that religion or religious belief reacts automatically to the machine as to a threat to its very existence but that whatever seeks, however naturally, to protect its untouched or untouchable integrity—the human organism, indeed, life itself—resorts to a similar mechanism: "In the end it is not only religion but life itself that is autoimmune, life itself that reacts in an automatic, machinal way."[18] The consequence of that is far-reaching, for an organism or a life that reacts automatically, machinically, to the machine is a life that is giving itself over, and "realizing" that it is giving itself over to being technologized in its most natural place or moment, a life that cannot exist or survive without the very technology that seems to mean its destruction. It means, therefore, that "life itself . . . must now be thought in relation to the machine,"[19] or as Naas later writes, "no life before the machine."[20]

Even the most benign form of autoaffection—praying, or the joy of wheelbarrowing—exposes life to the autoimmune threat: every self-caress opens the structure of self-mutilation. The very forms of autonomy or self-regulation that life requires bring with them elements of the autocratic and arbitrary, and, as *Rogues* discusses, the same holds for the organism called a body politic. Autoimmunity is thus a mechanism from which, as *Politics of Friendship* tells us in one of the first uses of the word, "no region of being, *phúsis* or history would be exempt."[21]

Back around the time when he was leaving the room and dying in *The Animal That Therefore I Am,* Derrida already pointed to "some analogical or virtual relation" between autoimmunity and autobiography. Autobiography infects as much as it affects the self; it is as much at the risk of becoming autoimmunizing "as is every *autos,* every ipseity, every automatic, automobile, autonomous, auto-referential movement" (*Animal,* 47). Even in a straightforward sense, we can understand that the more a writer has recourse to explanation or self-exculpation (Rousseau would no doubt be a prime example), the more he or she also runs the risk of self-indictment, being interpreted either as one who doth protest too much or as neurotic or paranoid. "Nothing risks becoming more poisonous than an autobiography," writes Derrida, "poisonous for itself in the first place, auto-infectious for the presumed signatory who is so auto-affected" (47). In his own case, he explains, there exists a proportional relation between autobiographical impulse and the proliferation of animals in his writing: "my animal figures multiply, gain in insistence and visibility, become active, swarm, mobilize and get motivated, move and become moved all the more as my texts become more explicitly autobiographical, are more often uttered in the first person" (35). Perhaps first among those animal figures is the animal-machine of *Of Spirit*: "This animal-machine has a family resemblance with the virus that obsesses, not to say invades everything I write. Neither animal nor nonanimal, organic or inorganic, living or dead, this potential invader is *like* a computer virus. It is lodged in a processor of writing, reading and interpretation" (39). Do we take that to mean that all his texts become animal-machine and hence proportionately autobiographical—that the animal-become-machine that is autobiography also invades everything he writes? And, to the extent that we can presume that that viral life-form will also go by the name of autoimmunity, does it not perhaps suggest that autobiography, in both the specific sense of self-writing and the general sense of autodeixis that I have been developing here, is a type of paradigm for autoimmune life, that autobiography is not just in an analogical or virtual relation to autoimmunity, but that *it is the very autoimmunity within writing*?

The "I" that says "I can die" is a subject in autoimmunity. In repeating itself, remobilizing itself, from physical context to physical context (pages 10 and 160), and from semantic context to semantic context ("I can die," "I can live dead," "I can die alive," "I can sur-vive"), "I" is performing that very automatization of itself in the world, setting in train a machine that utters itself by itself, both without itself and within itself. Anyone, finally, can say "I can die." Derrida will say it again, like clockwork, in his final interview published fewer than two months before his death, uncannily—but not so uncannily, for everything there repeats what he had been developing for many years—summarizing many of the themes elaborated earlier. Speaking of survival ("a question that has haunted me, literally *every instant* of my life") and of a legacy he can't foresee, he repeats, "I leave a piece of paper behind, I go away, I die."[22]

"I can die" is something everyone has to say, simply by living and being mortal. As Derrida suggested following Heidegger, and as we have been analyzing in detail here, unless one says it, nothing can really be *as such,* which makes a being strangely beholden to the possibility of being dead, not in the sense of recognizing its own mortality but in the sense of inhabiting the structure of death that comes from every other being having turned its back on it, having left the room, gone away, and died. That would mean that a being can only be in what I call dorsal space, the space that opens once another being has turned its back, left the room, or died. A being *is,* indeed, by virtue of inhabiting that dorsal space, by being behind the being that has left it behind in order that it might be. It *is* in the space of the unknown, of what cannot be known, for presumptive knowledge about how a being is is precisely what prevents a being from being as it is. Dorsal space, because it implies what is unknown, is also the space of surprise and of threat. In letting be by turning one's back, leaving the room, or dying, in declining to presume how a being is, one allows a being to invent what it is as something that precisely cannot be foreseen or controlled. It is the space of our relation—a necessarily ethical relation—to radical otherness such as that represented by the inanimate.

In letting be utterly, as it were, absolutely, to the extent of dying or leaving the room, one is in the first instance allowing oneself to be radically inscribed by death. Indeed, a radical ethics of letting be requires that level of self-abnegation or immolation beyond even the hospitality or being-hostage to which Levinas refers[23]—the suppression of one's every "vital design," as was suggested at the beginning of this chapter. It would require, in the final analysis—somewhat diverting or perverting Heidegger's terms—that one become less than the other's dog or some other animal, that one become more like a stone for the other. An inanimate thing is what one becomes for the other once one is dead, which leads to Derrida's provocative and moving definition of the other, in *Beast and Sovereign II*: "The others—what is that? Those, masculine and feminine, who might survive me . . . the other is what always might, one day, do something with me and my remains, make me into a thing, his or her thing [*faire de moi et de mes restes quelque chose, une chose, sa chose*]" (127). Such a definition, which leads back to the whole discussion of the automatizing thingness of survivance just developed, should also be put into correspondence with Derrida's discussion in volume 1 that turns around Lacan's apparent restriction of ethical obligation to one's fellow *(semblable)*:

> A principle of ethics or more radically of justice, in the most difficult sense, which I have attempted to oppose to right, to distinguish from right, is perhaps the obligation that engages my responsibility with respect to the most dissimilar [*le plus dissemblable*, the least "fellow"-like], the entirely other, precisely, the monstrously other, the unrecognizable other. The "unrecognizable" [*méconnaissable*] ... is the beginning of ethics. So long as there is recognizability and fellow, ethics is dormant. ... So long as it remains human, among men, ethics remains dogmatic, narcissistic, and not yet thinking. (*B&S I*, 108)

It is a logical step, obviously, from that call for ethical responsibility toward women, other races, and other species to other forms of life (e.g., plants) and to forms that we do not recognize as alive: "it is not enough to say that this unconditional ethical obligation, if there is one, binds me to the life of any living being in general. It also binds me ... to something non-living [*à du non-vivant*]" (110). In that sense, the logic of iterability would have to reach something of a terminal moment in the possibility of an ethics and politics of the inanimate. Simply by virtue of being repeatable, extracted from a supposed original context, the utterance enters into a relation of response, however irreducibly automatic or *reactional* it be, which opens the structure of livingdead remains and, by extension, of ethicopolitical responsibility toward the inanimate. The utterer—let's say the autobiographer—partakes of that relation and hence is immediately, from the beginning, rendered livingdead and bound by an ethical obligation to the remains he or she produces: in its most reductive form, that obligation inscribes a regime of truth(-telling) upon autobiography. Similarly, we can easily conceive of the ethical responsibility of the receiver of the utterance, bound to respect protocols concerning how to treat what is left to him or her: that tells us within what limits of respect or nonrespect a reading may take place. But by the same token—and it is here that the question becomes far more complex—the utterance responds to itself and so also functions within a type of ethical regime, one that is certainly unrecognizable, to the extent of defying the sense of the ethical or being easily dismissed as absurd, but that nevertheless remains within the logical gamut of the otherness that we accept as necessary for any ethical gesture to begin. If we can identify a type of becoming-inanimate on one side and the other of the ethical relation—allowing the other to be to the extent of extracting oneself absolutely, becoming a "dead" thing for the other—then it would have to be because the relation itself contains a structure of the inanimate. If recognizing a reactional automaticity in every response is the means to challenge human linguistic, communicative, or technological prerogative on the way to ethical treatment of the animal, then that would be because every life-form entails a question about the form of life up to and including what separates life from nonlife. The questions we have been dwelling on here answer not, in the final analysis, to some near or far-fetched conceptual speculation, to some ungrounded fantasmatic possibility, but instead to an unavoidable ethical obligation; as a consequence of them, Derrida insists, "one must inscribe death in the concept of life" (*B&S I,* 110). Beyond responsibility to the human other, to the animal other, to the earth, there is (even if not in the same terms) the requirement (even if it be motivated by the desire for our own continued well-being) that we come to logically coherent terms with our inanimate and inanimated other.

II
TRANSLATION

4

LIVING PUNCTUATIONS
Cixous and Celan

The previous chapter gives us to understand something of the complex operation by which the *autos* dies and lives in its own biographical, or any other, writing: how it dies in and as it writes but continues to haunt what remains; how, notwithstanding such a form of survival, the words themselves function only by being radically severed from their authoring instance. That rupture is precisely what allows them to live, and live on, as it were, anonymously, autonomously, and automatically, independently of their source. But if words are given life, and assured of survival in the ways just mentioned, can we conceive of their being nevertheless capable of dying? Do we accord the written word the capacity to end or die—borrowing Heidegger's distinction between the animal's *verenden* and *Dasein's sterben*—to simply perish, or rather experience its ownmost death? Or do we rather conceive of it in terms of the two fundamental categories of being: existence or nothingness, presence or oblivion? Can it be only as either legible mark or empty space? What is the status of what lies behind it or beyond it? What of its ashen remains? Or its lost fragments? And how does all that relate to what has come to be understood as the *trace*? In terms of the general thematic of this section, to be made more explicit in the following chapter, how does it *translate*?

Those questions obviously begin in terms of how the ontology of the written word is understood as its materiality. They might be considered specious were it not for various "abnormalities" that are readily ascribed to the verbal object, for example, the particular versions of finiteness that define a written text, the presumption that it has a beginning and an end, a beginning that is a formal incipit—such as publication—different from its genetic antecedents and an end that is either contrived by its author or left suspended, in the case of an unfinished text. That is to say, the surviving capacity of a written text that permits it to relive, potentially indefinitely, in every rereading functions in strong contrast with the finitude of its form, its formal beginning and end. Nevertheless, we tend to ascribe to that form a type of "organicity." We speak of it not only in mechanist terms—narrative arc or trajectory—but also in terms of rise and fall, or apogee and cadence, which evoke various life processes. But the analogies that enable comparison between the finitude of an organic life-form—the cycle of birth to death—and the structure of a written text, or between how the text lives on and, say, the memory of a dead person, are decidedly inaccurate. The text has not formally changed from one reading to another, whether those two readings be two minutes or two hundred years apart; and the formal "life-span" defined by its beginning and end is nothing like the actual state of animation that keeps it alive from one reading to the next. Hence, if the life of a text has no clear analogical relation to how we understand organic life, can we, conversely, still presume to call the usage of the word *life* as it relates to a written text metaphorical? Is there simply real, literal, organic life, on one hand, and a series of metaphorical extensions of that literality, loose figurative usages of the word, on the other? Or rather, doesn't life function through a variety of forms that never reduce to organic examples, however dominant and numerous the organic examples be?

The discussion in this chapter will concentrate on certain formal structural elements of the written text: the matter of what and where it is, and what and where it isn't. In what way do words end when they reach their final period or full stop, there where they leave off precisely in order to live on, given that their closure is also the opening to limitless commentary? Where is the end itself to be situated: in a final word, in a black punctuating mark, or in the white space that succeeds that mark? For the white space that represents the words' immediate context is indeed the locus of an unbounded juxtaposition: there is no limit to what that space can be made to contain. If, on one hand, the black letters of the text sustain the integrity that the text itself defines—determined by what its author has actually inscribed on the page—the white space around those letters, even within them, opens, on the other hand, an infinite and indeterminate set of supplementary recontextualizations. Yet the black letters can function only in contrast with their white "background," and it is difficult to maintain that those black inscriptions alone live and breathe in a given textual situation. If so, what status should be given to the blank space, there before the writing, there behind the writing, there beyond or after the writing? Should we conceive of it as the dead other of writing, or the assurance of its survival, or both?

Those seemingly marginal questions, I will argue, are tied to the highest ethical and political stakes of poetic, and, by extension, literary, expression. For I read the black-and-white relation between the "positive," manifest text and its blank spatial context as inseparable from the production, or the technological *poiesis* that determines all utterance and is thus inseparable from the whole vexed question of literary representation, precisely the relation of a text to what we call real life. Specifically, I contend that the relation between poetic expression and breathing, the play of inhalation and exhalation thanks to which we live and are able to express ourselves, in fact relies on its own (inanimate) interruption: a turning of the breath out of the breath occurs to inanimate the life that breathing sustains, and such a turning can be identified as a poetic function. In making that argument, I'll concentrate less on the words themselves that constitute what is expressed than on the spaces of rupture within that expression, spaces between or within the words, the periodic pauses that punctuate and so interrupt what is printed.[1]

The blank white page, or white space on the page, has of course a whole scriptural mythology: it is *incarnated,* given flesh and substance, but also made to grow abyssally into itself, by Mallarmé. By the time of his *Un coup de dés* (A throw of the dice), white space is no longer either the basis or background, the support or *subjectile,* for the printed word; it constitutes rather another type of poetic existence, another script or figure that conjoins with the black letters themselves and puts itself into play with them. On one hand, it speaks for itself, at once enunciating and contradicting the void; on the other hand, it is seen to invade the words and compose their letters, to provide a syntactic consistence and depth to match semantic consistence and depth. In that way, it constitutes a dorsal or *umbral* pictorialization that exceeds any opposition of figure and ground, and it represents a cosubstantial "agent" in the black-and-white fractal chiaroscuro by which writing comes to, and is sustained in, light, albeit as darkness.

Belgian artist Marcel Broodthaers created a 1969 version of Mallarmé's poem that reproduces it, in every way, we might say, except for the words themselves, which are replaced by solid black bars (Figure 1).[2] It is as if the whole contour of the lines of the poem had been redacted, for security reasons, with a black felt pen, leaving for our eyes only white space and solid blocs of print. It is one piece within a multimedia series of plastic and pictorial rewritings of Mallarmé's poem that Broodthaers produced in mimicry of the former's Book.[3] One is led to ask where Mallarmé's words have gone, whether they remain drowned or entombed under the solid back lines, whether and how they survive there: by means of memory for those who know them by heart, indeed, like the still beating heart of the poem buried alive below the now uniform printed surface? Or have they been erased, suffocated, and died, their scriptural and legible form made to succumb to a pure graphic force, the pictorial weight of undifferentiated solid black print? On further reflection, however, it becomes clear that what has in fact disappeared in the difference between Mallarmé's and Broodthaers's throw of the verbal dice, what has been subtracted from the one to produce the other, is not the black print that previously spelled out the letters, for that remains—virtually and actually—in the amalgamated black bars. Rather, it is the white space within and around the letters, their immediate white environment, that has been blacked out, which again reinforces the fundamental positivity of such blank whiteness in the monochrome play that constitutes a written text.

The scriptural deployment of Mallarmé's *Un coup de dés* can be seen to represent poetry's ultimate rupture from its oral origins. If we presume poetry in general to have evolved from song, to have begun as the intoning of a chant, then the form of its deployment on the page, as written text, will have been determined by the rhythms of the voice. That is what is called, precisely, versification: the arrangement in lines *turned* to suit an oral recitation. Within that schema, the more or less regular insertions of white space that begin to invade the page—between one line and the next, or between one stanza and the next, but also the caesurae within or between words—are understood to derive from, and replace, forms of punctuation that are themselves transcriptions of the human breath. No doubt the massive institutionalization of the alexandrine, in the French tradition, obeys the same pneumatic necessity. Racine's or Corneille's lines

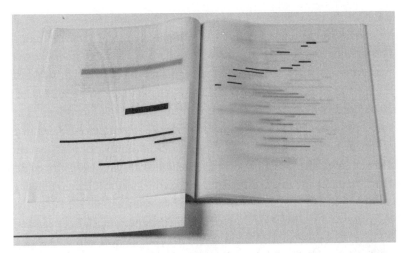

FIGURE 1. Poetry as pure *graphism*. Marcel Broodthaers, *Un coup de dés jamais n'abolira le hasard*, 1969. Print on paper. Musées royaux des Beaux-Arts de Belgique, Brussels. Copyright 2014 The Estate of Marcel Broodthaers/Artist Rights Society (ARS), New York/SABAM, Brussels. Reproduced with permission.

come like the calm, or fitful, breathing of the characters who speak them: "Ariane, ma soeur, de quelle amour blessée, Vous mourûtes aux bords où vous fûtes laissée" (Ariane, my sister, by what Love were you wounded and left to die alone upon that shore).[4]

Once verse becomes free, however, freed from that metric regularity, it also emancipates itself from the voice, or at least from an oral delivery that respects the relation between voice and breath: "l'acte vide / abruptement qui sinon / par son mensonge / eût fondé / la perdition / dans ces parages / du vague / en quoi toute réalité se dissout" (the empty act / abruptly which otherwise / by its falsehood / would have founded / perdition / in these latitudes / of indeterminate waves / in which all reality dissolves).[5] Free verse no longer promotes a voice that allows breathing to function undisturbed, and increasingly, or in certain instances, it will no longer promote voice at all (which is not to say that the history of poetry since Mallarmé does not continue to deal with the question of orality).

The white space of the poem is thus to be read, in the first instance, as the linear transcription of pauses and inhalations that prose, presumed not to be for reciting, treats much more liberally. It is precisely such a poetic line, with its vocal and pneumatic rhythms, that comes to be interfered with by the modernist experiment, as announced by Mallarmé's breaking news—"*on a touché au vers*"—delivered in Oxford in 1894.[6] However, the interference, "touching," or defilement of the line of verse that he spoke of will have been taking place since at least the prose poems of Baudelaire, and indeed within the latter's particular extenuations or exhaustions of the alexandrine itself. Already by the middle of the nineteenth century, the linear narrative of poetic genesis that ascribed it to an oral priority was being interfered with and called into question. If Baudelaire could destroy the alexandrine at the same time as he perfected it, less by abusing its subject matter, by using the pure aesthetic refinement of the line as the context for multiple indecencies, than conversely, by causing a naked carnality to syncopate and interrupt the flow of a regular and untroubled rhythmic breathing, it was because the line of verse harbored within itself pauses, breaks, or spaces that could be exploited beyond rhythmic orthodoxy. It meant that the line was punctuated not just by the white space at its end but in its very structural formation, that it was always as much a matter of catching the breath as of enabling it.

From that perspective, white space comes into play as a fundamental poetic principle that operates irrespective of vocal constraint. The formal *poeticization* of language is in the first instance its spatialization, made explicit by, but not limited to, the reduction or absence of punctuation. The very fact of the line—its interruption, rather—inscribes white space as the material signifier of poetry, such that it is no longer simply a function of orality, no longer simply the signifier of silence or inhalation, no longer what remains when writing breaks off, but is instead part of the writing. The invasive white space of *Un coup de dés* is thus less the expansion of "unused" scriptural space—even a whole stormy ocean of it—than the other face of the words themselves, words whose life no longer seems to derive from the breathing voice that emits them but instead comes from the animation of inanimate marks, from something that pulsates otherwise in the conjunctive play of blank and blackened space.

Gertrude Stein once wrote, "When I first began writing, I felt that writing should go on, I still do feel that it should go on but when I first began writing I was completely possessed by the necessity that writing should go on and if writing should go on what had commas and semi-colons to do with it, what had commas to do with it, what had periods to do with it what had small letters and capitals to do with writing going on which was at the time the most profound need I had in connection with writing."[7] The writing that goes on is here an unstoppable flux, a writing impelled by its own expressive force to exceed the strictures of punctuation or syntactical closure. In terms of what I was describing earlier, it would be understood as a type of pure pneumatics, writing as prolonged exhalation, an overwriting of both punctuation and the blank space that poetry had introduced to replace that punctuation. Not that such a pneumatic writing is without poetic force, not that it doesn't involve a poeticization worthy of the name, but it is a writing that both erases and reinscribes the oral genesis of poetry, replacing regular metric breathing with a type of breathless aspirational overflow.

We find a very different example in the writing of Hélène Cixous. Early in *First Days of the Year*, a text from 1990, her third-person "author," who is in conversation or competition with a first-person narrator, is discussing how she has spent thirty years trying to write a book. She recalls Paul Celan's poem "Cello-Einsatz," from his 1967 collection *Atemwende*, and brings to mind more generally her relation to Celan's writing as it concerns memory and suffering, tears for something lost that veil the eyes, the sacrifice of one pain for another, longing for a language that one possesses incompletely. Celan's writing unfailingly makes Cixous's author cry and produces an amalgam of poem, grief, and sobbing, doing so specifically by means of its interrupted breath, the pant or gasp of its rhythm:

> Just as the cello had been created to moan the animal music of our entrails and the oboe to give wings to the triumphal moods of our adolescences, so a Celan had been created for singing, his mouth full of earth, under the century's cleaver, under the pickax, the unique little slip of fleshy paper that will have succeeded in escaping the shovel of the Apocalypse. . . . And she had wept for several years with one of Celan's tears. A God had provided for her need to weep by inventing Celan, the poet with the name in reverse, the poet who started out being called Ancel, then stopped being called Ancel, then called himself Celan, and thus had emerged from the forgetting to which he had been consigned, *by calling himself contrarily,*

and behold him standing on the silent soil, his chest full of cello boughs.
Only thus are we able to advance, by beginning at the end, death first,
life next, teetering life next [*ensuite la vie chancelante*], teetering so [*si chancelante*], chance so [*si chance*], concealing/"celaning" so [*si celante*],
    Mused she, the author, trembling
    Sensing it was useless to deny the event.[8]

I shall ignore here, woefully, the extraordinary resourceful connections and the powerful articulation that Cixous's narrator-(sometime)author makes in these lines between a scapegoated Celan and the repressed event of her own father's death, indeed, the whole conceit of her as yet unwritten book that will emerge, however fitfully, once "these years of passion by proxy" (*First Days*, 10) that a god called Celan provides have come to an end. It would also be necessary, were this discussion to give the space of commentary that Cixous's passage merits, to relate the particularities of Celan's poetic singing with a mouth full of earth to what has just been developed concerning a poetry divorced from its orality. Some of those resources and articulations will return, either explicitly or implicitly, in what follows. Instead, I wish to grapple in the first instance with the strange play of punctuation and spacing by means of which Cixous, under the express influence of a Celan-made-verb, allows her prose to become poetry. The portmanteau of chance and Celan, packed into the participial form of the verb *chanceler* (to teeter), is followed first by a comma, as in normal prose. But, following the comma, that line breaks off in mid-line, giving the appearance of a poetic format that is reinforced by the following indented line, which also breaks before reaching the right hand margin, this time without any punctuation ("Songeait-elle, l'auteur, tremblante"). Another indented line follows, but it is printed all the way to the right margin and ends with a period, ostensibly signaling a return to prose format (Figure 2).

What of it, one might well ask? In the first place, one can track similar lapses in the use of punctuation in any number of other texts by Cixous. An inventory of such lapses would include, to give but a tiny chronological sampling, (1) a chapter toward the middle of *Dedans* (1969) that ends with a colon ("Un jour, il me depose ; ainsi :"); (2) forward slashes at the beginning or end of a paragraph in *Le Troisième corps* (1970), in

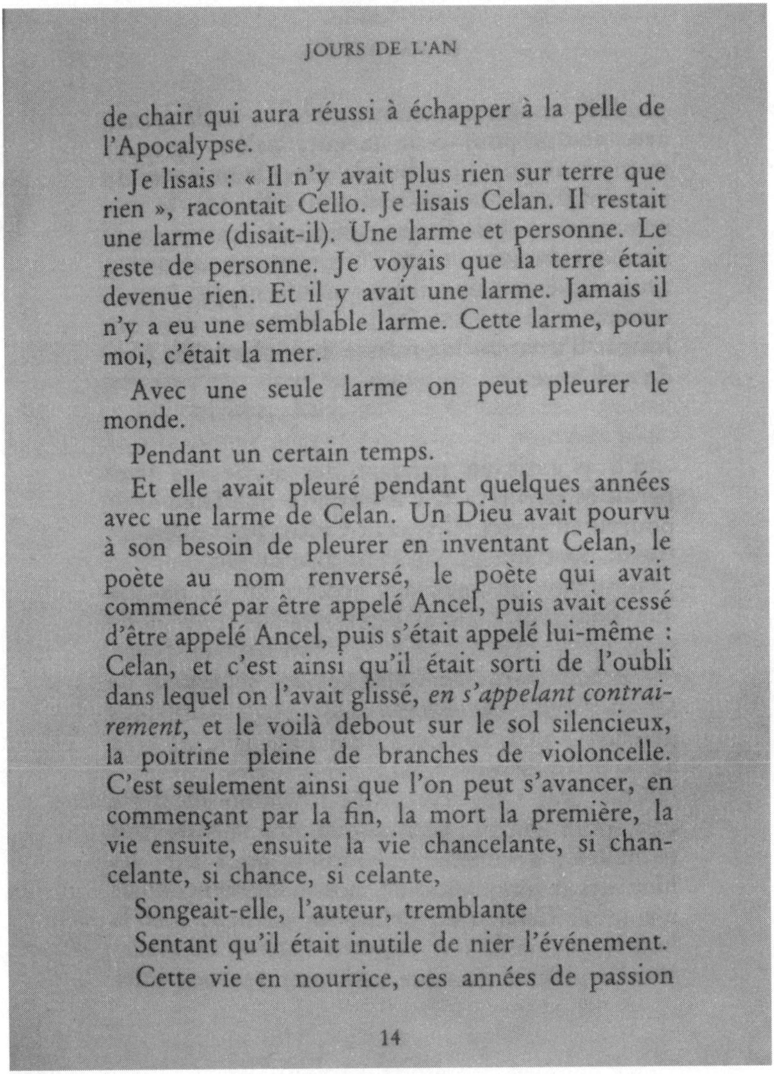

JOURS DE L'AN

de chair qui aura réussi à échapper à la pelle de
l'Apocalypse.
  Je lisais : « Il n'y avait plus rien sur terre que
rien », racontait Cello. Je lisais Celan. Il restait
une larme (disait-il). Une larme et personne. Le
reste de personne. Je voyais que la terre était
devenue rien. Et il y avait une larme. Jamais il
n'y a eu une semblable larme. Cette larme, pour
moi, c'était la mer.
  Avec une seule larme on peut pleurer le
monde.
  Pendant un certain temps.
  Et elle avait pleuré pendant quelques années
avec une larme de Celan. Un Dieu avait pourvu
à son besoin de pleurer en inventant Celan, le
poète au nom renversé, le poète qui avait
commencé par être appelé Ancel, puis avait cessé
d'être appelé Ancel, puis s'était appelé lui-même :
Celan, et c'est ainsi qu'il était sorti de l'oubli
dans lequel on l'avait glissé, *en s'appelant contrai-
rement*, et le voilà debout sur le sol silencieux,
la poitrine pleine de branches de violoncelle.
C'est seulement ainsi que l'on peut s'avancer, en
commençant par la fin, la mort la première, la
vie ensuite, ensuite la vie chancelante, si chancelante, si chance, si celante,
  Songeait-elle, l'auteur, tremblante
  Sentant qu'il était inutile de nier l'événement.
  Cette vie en nourrice, ces années de passion

14

FIGURE 2. Cixous's idiosyncratic punctuation and spacing. Photographic reproduction of Hélène Cixous, *Jours de l'an*, page 14.

the second case following a period; (3) in *Angst* (1977), paragraphs that
end with an ellipsis, a comma, or a period plus a dash; (4) in *First Days
of the Year*, the first two paragraphs that end with commas (*First Days*,
3) and the unwritten book referred to by means of a colon between two

dashes (*First Days,* 5); and (5) from the first page of *Insister,* the fragments of a conversation that end, seemingly indiscriminately, with or without a period.[9] These random examples lead one to question the logic of Cixous's strategy vis-à-vis punctuation—a question for which I have no coherent response other than to point to the obvious fact of a certain hesitancy before the syntactic closure represented by a period or full stop, or else a stubborn resistance to grammatical law—but the practice clearly demonstrates a desire here and there to impose or maintain a formal suspension of the sentence.

In the second place, the liberties Cixous takes with the codes of punctuation point to the problematic status of diacritical marks within writing conceived of as a transcription of spoken language. A period clearly does not represent a spoken sound but signals rather the absence of sound or a type of pause, and the precise taxonomy of such "unsounded" pauses—period, comma, semicolon, colon, dash—would be difficult to establish. For in each case, it is a matter less of a quantifiable and hierarchical amount of silence than of a pause that functions as an element within an intonational sequence extending over an ill-defined syntagmatic time and space. That is all the more so in the case of question or exclamation marks. Where punctuation does seem to represent what we might call a "neutral" pause or silence, as in the case of a period or comma, its neutrality is belied by the extent to which those diacritical marks have in fact been divorced from oral transcription in general and pressed into service on behalf of the writing itself, dividing it and restructuring it, imposing different hierarchical levels upon it, in ways that no longer answer to any purely oral function. Not only is the spoken version of a written text unable to account for all the differences among punctuation marks—distinctions among comma, dash, and semicolon, for example—but such marks often require speech to resort to periphrastic glosses to convey their meaning.

In the third place, therefore, punctuation comes to function as a form of *parenthetization,* the opening of a putatively subordinate space that may always proliferate and threaten the status of the metasyntactic sequence that it interrupts. Its presence gives a certain punctual visuality to that space, but the absence of it gives that opening a different kind of explicitness, a sequential visual spatiality similar to the writing. And at the same time, the blank space makes the opening abyssal: it is reducible to a finite set of letters on one level but is unquantifiable both in breadth and depth on another level. How indeed does one measure or confine blank space once it is broader than a single character? Conversely, the absence of, say, a period is an absence of words: words that could have been used to represent it (simply "no period appears," "blank space," but also all the words constituting this or any other potential discussion of it) and any words whatsoever that could fill the opened space, still more words, contextually nonsaturable, anything that could be made to fit. For finally, the suppression of all punctuation, such as briefly occurs in the example from *Jours de l'an*, extends the question of Cixous's logic and strategy at the same time as it asks in more general terms what is at stake where no printed mark imposes closure: is it simply, or even complicatedly, the poeticization of prose that I have discussed, or does it signal a radical openness that destabilizes not just a given sentence but every mark, making every mark a random graphic *trait* and throwing terminally awry the controlled relation between printed mark and blank space that allows a written text to function and to signify? Furthermore, as I shall be asking later, for this is where my whole discussion is heading, how might the question of an interrupting or terminating mark such as a period, its presence or absence, be construed within the context of the life, death, and survival of writing?

Derrida will have parsed elements of those questions in his own comments on Cixous's use of punctuation. Referring specifically to her text *OR*, he notes that the punctuation is sometimes removed, and he adds in parentheses,

> Nobody knows better how to punctuate, in my opinion, and to punctuate is to write, nobody knows better how to remove or spirit away [*enlever*] punctuation than Hélène Cixous, yes, spirit away—whether it be marked or not, her punctuation is spirited (away) [*enlevée*]—and one should not allow oneself to speak of *OR*, as of all her other books, without first dealing with punctuation: before the letter, there is her punctuation, which is like the first silent letter of all her letters and the gear change for all her different speeds.[10]

Shortly thereafter, he adds, "Nobody, need I repeat, can compete with her when it comes to a genius for, and meticulous calculation of, punctuation— which is, one can never say this enough, the heart and as it were the living breath, the very lungs of the writing. Here punctuation removes itself or gets spirited (away) *(s'enlève elle-même)* by a punctual *depunctuation* of its very breath, in other words its life, its rhythm, its time, and precisely, its speed."[11]

Now, by one reading, Derrida is here reinforcing the relation between punctuation and oral in/exhalation that I called into question earlier: punctuation is said to be the heart, lungs, and living breath of writing. By another reading, he is acknowledging that the absence of punctuation—whose paradigm I shall henceforth take to be the period or full stop—interrupts (punctually *depunctuates*) that breathing, altering its time, rhythm, speed, and, indeed, life. Indeed, it is my hypothesis here that it is punctuation itself, precisely as writing, that depunctuates the breath of language, punctuation itself and not just its absence; that punctuation, in fact everything that turns around, within, and behind the smallest point or spot of writing such as the period, its minimal but not necessarily least enunciative gesture, concerns less the heart, lungs, or living breath of writing than its syncope or caught breath and derives rather from something like a transplanted, prosthetic, or artificial form, or organ of living.

Again, that would seem to be contradicted by Derrida's reference to punctuation that gets "spirited (away)," punctuation that, in being simply removed *(enlevée)*, also undergoes a type of uplifting or sublation *(relevée)*. For although the sense of "spirit" that the translators have used is not explicit in the French, *enlever* does imply both that punctuation is carried off and that it levitates. Derrida's idea would be that Cixous renders language more ethereal, performing upon it an interference that has something magical about it. Indeed, he will later refer to "the magic of what, by a stroke of writing, does the impossible," going on to admit that magic "is a tempting word, a dangerous temptation," one that risks "a phantasm of animist and infantile omnipotence," an occult obscurantism.[12] Now, magical animism is certainly not the sense of the technological inanimation that I am promoting here. Prosthetic life comes to be, as it were, out of logical necessity, as a constituent element of life, not as the result of some occult conjuration. The relation of black print to white space that I am trying to describe beyond a contrast between two materialities, as the mobilizing of an *in*animat*ed* trace within the process whereby inspired breath is formed as concrete expression, has nothing magic, nothing invisible or ineffable, about it. Magic would precisely be the mythological investment of a very traditional conception of technology, one based on a reductive creationist sense of invention, the *coup de génie* or stroke of brilliance that is the other side of a pure causal mechanism, the voluntarist excess that inspires an inventor to take things just one step further or in a slightly different direction to produce the new.

Nevertheless, it is true that invention must carry with it something irreducibly other, unforeseeable, unthinkable, and unaccountable—what I have elsewhere called the dorsal surprise[13]—if it is to have any sense at all, and precisely if technology is not to be reduced to a manipulation of the material solely within the grasp of the human in particular and the animal in general. Rather than to delegitimize any reference to magic, therefore, what might be more important as a strategy is to insist on the possibility of a magic that rejects occult obscurantism precisely by being mobilized within the network of the technological. That is Derrida's strategy in favor of Cixous, via Freud's concept of an originary *Belebtheit,* to insist on there being "no longer any contradiction between the experience of magic and the experimentation of the most objective techno-science. . . . No distinction then can hold any longer between magic and technique, faith and knowledge, and so on."[14] Hence punctuation in Cixous would have to be understood to be both removed, as if surgically (*enlever* has that sense of ablation), and spirited away, as if magically. And what it leaves in its absence would have to be both pure ether, endless agnostic speculation or gnostic conjuration, and a technological far side or other side, something like the life of blank matter, the other side of the enunciative scriptural machine. If letters represent the expressive exhalation, and punctuation the revivifying inhalation that reenables the flow of written discourse, then missing punctuation allows that discourse on one hand to melt into the thin air of poetic ineffability. But on the other hand, missing punctuation concentrates in the white space that it opens up an arena of blankness that is no longer subordinated as substratum to the writing nor indeed that functions as the organic support and basis for the endless accumulation of the artificial appendages of writing. It is instead the necessary and constitutive prosthetic potential of writing itself, the site of articulationality, the space where white articulates with black, the profile that is traced of the very technological substance or livingdead matter of signification itself.

My reference to the surgical removal of punctuation is prompted by another passage that Derrida lights on elsewhere in Cixous, from *Dream I Tell You*, where the dreamwork is compared to an "*invisible laser scalpel*" that slips "*between the letters: t, u, t'es eu, tu,* {you've been had} *next between the signifieds Siamese twinned by homonymy: tu es tu* {you are you} *that's why, étant* tu {being you, having remained silent}, *tu ne* plus *te taire* {you can no longer remain silent}. *As for the bistouri* {scalpel}, *il bisse tout ris* {repeats, echoes, all laughter}."[15] Now it is evident that a translation cannot hope to convey all the versions of this free play of the signifier such as might operate in a dream. As soon as language loses its normal differentiating markers, particularly those—like punctuation, or phonemes and sounded letters—that play between its oral and written forms, it reverts to a type of indistinct and unarticulated babble, the babbling brook or churning torrent of a *flumen orationis* or unchecked flow of discourse. That produces two effects: one no longer knows where it begins or ends, borne along by the machinic automatism of a limitless expressivity, and language begins to sound as much like incantation as enunciation, as if it were repeating the spellbinding power of a conjuration. Cixous's punning indulgence reveals both the chaotic flux of such homonymic and other associations once the syntactical constraints of normal speech have been removed, producing something like an indiscriminate white vortex of sense, and the automatism of such wordplay, its machinic-like autogeneration. Rather than effecting a return to pure vocality, the "play" of the signifier engages in fact the automatism of iterability. Such iterability begins to be heard (even though it has been working from the first trace) in the differential repetitions of assonance and alliteration that a language consisting of a limited number of linguistic elements cannot avoid, and it is thanks to such repetitions that linguistic signification will never reduce to semantic difference without also engaging a whole range of poetic or musical effects. The obvious is perhaps not sufficiently, or not sufficiently frequently, registered here: for all that poetry gives us in terms of the repetitions of sonic signifiers, for all that it makes language sing, it also tethers it to the grinding predictability of a rhythmic or sonic machine.

But the logic of iterability, and thus the incantatory repetitivity of language, does not begin with a given voluntarist or even unconscious instance of wordplay. Because it is at work in every trace, and in every utterance, every utterance constitutes or potentializes a tautological repetition of itself. As soon as the signifier is deployed sufficiently to reveal its own depth, thickness, or consistency, it opens the space of its own internal mirroring. Every phoneme, syllable, or word—indeed, every blank space that utters, as we shall shortly discuss—becomes thus a homonym of itself, no longer being enunciated a single time but instead resounding within the echo chamber of its own space. *Tu,* therefore, before being both personal pronoun and past participle, and before being disseminated through the spelling of its individual letters, *t/u,* would be plural by definition, at least two times *tu* within each of its avatars. Simply by means of being spaced, *tu* would always be *tutu,* pronounced simultaneously.

Interrupting the automatic homonymic flux of poetic or oneiric language is an arbitrary operation: "I'd better stop," Cixous writes immediately following the excess just quoted. "I don't want to make the hackles rise on those who get too quickly scared off by the philosophical and philosophicomical resources of language."[16] But that decision is not in strict opposition to the excessive play that we have just witnessed, which was itself already a contrived, artificial intervention within language, which is itself, of course, a contrived, artificial invention, and which involves often complicated mechanical maneuvers so as to prevent its elements from getting confused one with the other. Such maneuvers begin at the most local level with etymological and syntactic restrictions and avoidances and go all the way to the standards of taste that are invoked against low forms of wit like punning, to prohibit or discourage it in formal discourse while permitting or encouraging it in poetry. Whereas dream provides an alibi for the play of *t* and *u,* and of *tu* (second person pronoun) and *tu* (past participle of *taire,* to be silent), the syllabic separations of *bistouri* into *bisse tout ris* are a fact of Cixous's allowing the logic of oneiric signification to invade her own waking discourse and thus call for, or at least raise the possibility of, some form of self-censorship.

However, when she says she'd better stop, Cixous will have in fact just launched a whole other chain reaction of signifiers. The verb *bisser* (L. *bis*) itself refers to a supplement or repetition coming at the end of a performance (when the audience calls for an encore, they shout, "Un bis!"), which means that the fine laser scalpel that removes the delineations between sounds and letters in order to have them freely signify also produces repeat performances of the same. It does that to everything *(tout),* provoking or exhorting laughter *(ris),* but also—who knows?—having signifiers multiply and expand like grains of rice *(ris).* Thus the encore performance of one *tu* being recited in a recital as another *tu* has the power to continue even as it is being suppressed or silenced, for example, by means of a blank space, a period, or the absence of a period. Indeed, in her text—and it is here that we return to questions of punctuation— Cixous performs that silencing four times, by means of four different graphic interventions: by means of a period, a new line, an indent, and the announcement that she will stop *(J'arrête).* It is as if she has punned her stopping via four successive graphic repetitions, each of the first three being increasingly "silent"—a dot, a form of elision, a blank space—while the fourth fades into the insignificance of its own pleonastic redundancy. In phenomenological terms, it might be argued that the blank space of the indent best performs the demurral or reticence that is being expressed; it is the blank space that is most eloquently silent, such that to suppress the threat of unlimited play of the signifier, one has no better recourse finally than empty white space, the blank space from which nothing more emerges to break up the flow of ordinary discourse. But it is also a blank space that represents its own form of flux as well as a type of abyss into which all those actual and potential signifiers in play have fallen. It should not escape notice that the word *bistouri* begins with a *b* and that if we were to allow that word to break up as in a dream or a poem and have its elements signify otherwise, one of the first, if not the first, such element to emerge, at least somewhere between French and English, would be a form of another one of Cixous's idiosyncratic verbs, namely, *béer,* "to gape," like a flabbergasted or silenced open mouth *(bouche bée)* that is another type of abyss, the simultaneous silence of nothing to say and too much to say. Then the scalpel might have begun its free fall from a dissecting table at the most critical moment of the operation, catching in the moving parts of a sewing machine running without interruption and tirelessly weaving some oceanic textile.[17]

Homonymic or poetic wordplay thus opens the structure of mechanistic sonic repetitions that threaten to reduce language to an irrepressible babble, or to the silence of incomprehensibility, as if signification were to disappear into the vortex of white space. In producing competing senses within the same textual space, the homonym creates a confusion that constitutes something of a graphic erasure, such as in Broodthaers's Mallarmé, but that erasure, or reduction of what is written to a play of black and white, nevertheless engages, even as it overstrikes them, both the positive and negative valences of written form. In another sense, however, the homonym, in doubling or pluralizing the signifier—that is, in explicitly performing the reiteration or iterability that structures every enunciation and every trace—inscribes a simultaneity that, as it were, *thickens* not just the black lines we call letters but also their constituent white space(s). The doubling of *tu* as both "you" and "silenced" takes place, of course, as a spatial simultaneity—both senses are recognized within the same graphic traits—but according to another logic that simultaneity is an impossibility, for in any given reading of the syntagm, only one or the other of the homonymic possibilities is valid. In order that the wordplay function, there has to occur either an infinitely rapid substitution of one possibility for the other or an imperceptible doubling of the writing, as if one *tu* were to be separated from the other, which "appears" behind it, by a white space that is both infinitesimal (no space at all) and infinite (sufficient to produce absolute difference). That sort of necessary impossibility makes for what I have described elsewhere as a technology of language functioning faster than the speed of light, of a single sensory-semiotic impulse that travels *at the same time* in two different directions.[18] However, in the context of white space versus black trait that we are examining here, it should also be understood as a technomagical destabilization of the letter that produces a strangely other materiality, a new relation of black to white, other than the traditional figure–ground relation, but also other than the "mobilization" of white space within, around, and behind the letters that I have been relating here to the disappearance of marks of punctuation. This technomagical materiality would have to be imagined, rather, by according a spatiality, and a chromatic valence, to the white, nondetachable underside of the letter, having that underside space exist sufficiently to separate the two (or more) elements of the homonym, a white tain applied to the letters allowing them to mirror themselves within themselves, comparable to the slash that we insert between them when they are spelled out syntagmatically.

In his reading of Cixous, Derrida has once again sought to theorize this phenomenon of impossible temporal and spatial instantaneity, analyzing it in terms of both force—to which I shall return in conclusion—and, in relation to homonymic play, speed:

> Speed should change its name because it operates this rhythmic or spatio-temporal *displacement* [in writing] only by beginning with *replacement.* Before displacing, it replaces. If it displaces so quickly, it is because it replaces. That is why it is infinite, or absolute, like an acceleration that goes faster than speed: even before moving and being able to move [*pouvoir mouvoir*], it replaces, it substitutes, it puts in the place of (one address for another, a word, a phoneme, a grapheme for another, one meaning for another).... Absolute speed, the speed that absolutely economizes on speed, is first of all the relation to itself as relation to the other of a metonymy or a homonymy that *replaces* a noun, a mark, the address or the meaning of a phoneme, of a syllable, or of a grapheme, etc., instantly, at once, without delay. Replaces them on the spot [*sur place*], at once [*sur l'heure*], and forthwith [*sur-le-champ*].[19]

The homonym destabilizes the signifier in its relation to itself as singular and in relation to a given signified. It displaces itself onto its other, not simply onto its other signified (*tu* displaced from personal pronoun to past participle) but onto the other relation of signifier to signified (*tu* as "you" displaced onto *tu* as "silenced"). The displacement is also, therefore, a replacement or substitution, a *two-sound monte* performed at infinite speed: it is as if you were hearing or seeing one and the other, one for the other, in the same place and at the same time.

Now it might be countered, throughout what I have been advancing here, that homonymy is in the first instance a fact of spoken language, a confusion deriving from a single sound's being required to perform double duty, a confusion heard rather than seen, especially because, in the case of impurely phonetic languages, the same sound may be written differently. But in calling attention here to a poetic *graphemization* of the word that divorces it from its oral form, and in emphasizing in particular the radical divorce of written from oral language that takes place by means of punctuation, more radically still by means of an absence of punctuation, I am arguing for something, as it were, beyond the fact of the incommensurate relation of spoken to written and beyond the difficulty, if not impossibility, of representing silence in graphic form. I am arguing for a homonymic "materiality" that both invokes the silence that enables sounds to be differentiated one from the other and dissolves that silence into an abyss of blankness that no longer answers to any recognizable function of representability. Such a blankness is at once the absence of any mark and the space that gives the very density of every mark—both the whiteness that figures the consistency of writing and the invisible reflecting surface by means of which a homonymic doubling comes to divide such consistency, having it fragment and dissolve.

If we were able to conceive of such a blankness, we might begin to envisage a homonymy beyond aural signification, for example, a homonymy that functions through the punctuation mark, say, a period. The American English word *period,* differently from the word *point* in French, already calls forth complex homonymies, referring both to an interval or duration of linguistic utterance, such as the sentence, and the mark or dot that terminates such an utterance. But what of the mark itself, the mark before its name? The " . " silently utters the end, terminal, or stop. It is understood, as I pointed out earlier, to represent the silence of a pause, but that would be in graphic contrast to the silence represented by blank space such as surrounds words arranged prosodically. Yet I am arguing for that graphic contrast itself to be distinguished from the white "silence" that both supports and invades the letters themselves, a "silence" that is in turn different from the blank space that comes into play—as it were, at the original poematic moment—when such a mark is withdrawn or refused. At least four different versions of white silence are therefore to be accounted for: period, non-/less-punctuated poetic spacing, the blankness that is a constituent element of what is enunciated, and the abyssal blankness that intervenes when the period is judiciously withheld in prose writing.

On that basis, how is a normal conception of materiality within a normal conception of oppositional difference to account for the homonymic or tautological sameness of a mark and a blank space? How can " . " and " " be homonyms? For that is indeed what is suggested by the multiple stops in the example from *Dream I Tell You* (period, new line, indent, utterance)? Or would the whiteness of an absent or disappeared dot be rather, as I have already tried to describe, the other of that period, its reverse side, as it were, the dot's *bright blind spot,* or again, the tain that precisely allows it to reflect itself homonymically, tautologically. We are faced with the double impossibility of a white underside to a black point and/or of a white tain that somehow blocks white, diaphanous, translucent light, transforming the mark in the course of its transmission to have it repeat itself each time differently, replacing itself, but each time with less focus and more infinitesimally, by means of progressive infinite regress toward its disappearance.

The question, then, is how to *auralize* or visualize such different forms of silence as they descend into that abyssal, resounding blankness, into what might be considered a purely homonymic blankness, or indeed, following Derrida, to try to determine the speed of their substitutability. For it would be a speed clocked at two times absolute speed or even infinite times absolute speed. That is to say, in the case of a disappeared period substituted by blankness, the substitution would have to take place, at least according to a logic of successive substitution that excludes simultaneity—perhaps the only possible temporal logic that takes account of displacement as replacement—in half the time of a replacement, hence at twice the speed. For instead of the disappearance of a graphic signifier followed by the arrival of its substitute, it would be a matter of imagining its departure only, its "simple" disappearance, therefore at double the absolute speed of a substitution involving another mark. But even supposing we were able to conceive of such an absurdity—a white spot, dot, point, or mark as other side of the manifest phenomenal appearance of such a mark and disappearing at two times absolute speed—even if we were able to account for the two homonymic possibilities of the other of the point that I have tried to describe, we would not for all that have begun to understand the corresponding sense of the blank homonym itself, the different substitutions of whiteness for whiteness, the homonymy of whiteness itself, of the effects of the tain and exponential reflection at work and play in that whiteness itself, of such pure abyssal whiteness. We would not for all that have understood what constitutes the "mark" in and of the very white absence of a mark. What would be the ontology or phenomenality of such a doubled whiteness, and according to what logic would we presume that it had none, neither existence nor phenomenological status, given the necessary structural iterability of the "manifest" mark, and given the

PLATE 1. The "angel of history," back turned to the future: Paul Klee, Swiss, 1879–1940, *Angelus Novus*, 1920. Oil transfer and watercolor on paper, 318 × 242 mm. Collection The Israel Museum, Jerusalem. Gift of Fania and Gershom Scholem, Jerusalem, John Herring, Marlene and Paul Herring, Jo-Carole and Ronald Lauder, New York. B87.0994. Photo copyright The Israel Museum, Jerusalem by Elie Posner. Reproduced with permission.

PLATE 2. The angel as prayer: Jean-François Millet, *L'Angelus,* between 1857 and 1859. Oil on canvas, 55.5 × 66 cm. (RF 1877) Paris, Musée d'Orsay, legacy of Alfred Chauchard for the Louvre in 1910. Reproduced with permission.

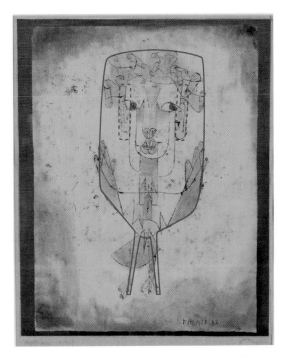

PLATE 3. Benjamin meets Millet meets Williams: "Wheelbarrow Angel."

PLATE 4. The thing itself: "Red Wheelbarrow."

PLATE 5. Corinne in "analysis": still from Godard's *Week-end.*

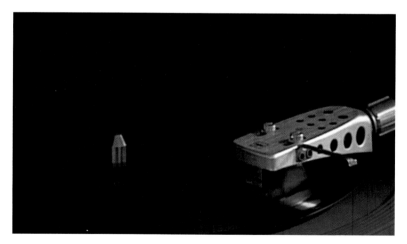

PLATE 6. Framing Bataille, second instance, shot 3 ("*Todesfuge*"): still from Godard's *In Praise of Love.*

engagement of white space in the constitution of such a mark that we have been dealing with here? More problematically or provocatively still, in terms of the overriding theme or concept I am calling inanimation, what would be its status as living(ness), as survival or *survivance*?

As I explained earlier, in *First Days of the Year*, it is Cixous's narrator's idiosyncratic experience of Celan's writing that causes her own words to teeter and fall out of prose into poetry, at least from the point of view of the explicit formal indicators constituted by spacing and punctuation. But poetry was already on her mind and in her experience. *Poem* was the word she had decided on as most appropriate to refer to the book she had been carrying with her for thirty years but hadn't managed to write, the book that *First Days of the Year* will, to some extent at least, become: "She had tried to figure out how to call this unwritten book . . . and, lacking a name for designating a thing that was not of this world, our own, the visible, had proposed to herself the word *poem*. . . . And what this book and a poem had in common was the physical sensation, the cardiac certainty, of their both belonging to a wholly other time" (*First Days*, 5–6). And it is Celan's poems in particular, "to which I would cling, weeping, as if to my own memories" (7), it was his poems, "Cello-Einsatz," among others, that gave her back her "lost tears" (7), allowing memory, and writing, to emerge, as in that poem, *von hinter dem Schmerz*, from behind the pain.[20] Before the word *poem* comes to her, however, the thing without a name that haunts her for thirty years, "this voice stronger than my voice, but which doesn't pronounce words," is represented by the three dots of an ellipsis: "This . . . , this poem, then" (5), and the same formulation is repeated a few paragraphs later (6).[21] That suggests at first reading that what will be called a poem before becoming a book, what is held back, waiting like tears behind the pain, unpronounceable, wordless, throbbing or beating with cardiac certainty and force in another time before emerging as poeticofictive writing, is an inspiration experienced like a holding of the breath: it is held in suspense before finding release as pneumatic expression, a palpitation, sob, or lachrymal overflowing:

> Eyes closed, reading the music with her heart's eyes, *Cello-Einsatz, von hinter dem Schmerz,* the poem between her breasts . . . taking pleasure

from each line she hadn't written. . . . Taking pleasure from each pain she hadn't had the chance to feel, but which, fortunately, had been felt—sung. And whose scansion, whose rattle she recognized. . . . At each reading, an unbearable sweetness upon hearing the words she might have moaned in that other life—with the two, with the four first accents—would burst into her chest, coming from the source of regrets, Cello-Einsatz, and for the hundredth time as for the first, the sob would rise, rise and with the poem's last accents break upon her heart's rock. (8)

On one hand, therefore, Cixous's narrator conveys there something like Stein's idea of creative poetic expression bursting through the pain with a type of pneumatic automatism, unable to be held back any longer. From that point of view, the ellipsis points mark what is not expressed, what cannot find words, what language hesitates to state, what it retains in an unuttered reserve that is necessarily a type of silence and an interruption of breathing. The fact that the poem rises and falls on her chest before bursting there, coupled with the words "scansion," "rattle [râle]," "accent," and "sob," reinforces a rhythmic exhalation that includes its own caesurae, a holding and releasing of the breath. But that pneumatic and enunciative hesitation, like a momentary aphasia—"what to say?" "how can I put it?"—also functions as an invitation to another, a place of encounter where a different interlocutor might make a paradigmatic insertion into the syntagmatic chain, offer to provide the word or idea that is lacking, however provisionally. That would be true for how Cixous has Celan help her utter her lost book, and its tale of loss, and also for how Celan's relation to the German language speaks to Cixous's own experience of that language ("mourning tongue, the cherished, inherited, hated tongue" [8]). In that way, Cixous's narrator's ellipsis-on-the-verge-of-poetry corresponds to what Celan will develop in his famous Meridian address concerning the poem as encounter with another and as turning of the breath, *Atemwende*.[22] *Atemwende*, it should be noted, is the title of the collection of poems that includes "Cello-Einsatz," and Celan's poetic scene is clearly staged by Cixous here. I shall return to it shortly.

On the other hand, none of that changes the fact of the ellipsis as series of graphic marks within the framework of what I have been developing, however prosaically or pedantically, concerning the relation between poetry and punctuation, or the absence of punctuation. What is held back, suspended in its wordlessness before becoming poem, is *scripturally* inscribed as a series of points—*points de suspension* is exactly what the ellipsis is called in French. If they represent the word that cannot yet be found, they also function as the sign of a potential or imminent fall into blankness, a spluttering that precedes an irreparable breakdown but that nevertheless gestures toward a different form of expressivity unhindered by syntactical law, a crumbling of the graphic façade that requires and inaugurates a more radical structural redesign of the writing, and of its relations between black-and-white traces. As we saw at the beginning, that is precisely what comes into effect two pages later, with the chance teetering into blankness via Celan's proper name become participle *(la vie chancelante, si chancelante, si chance, si celante)*: a new writing of a new silence.

We should therefore read in the "..." not just an emotive poetic encounter between Celan and Cixous but also an inanimation of bodies and discourses via the writing, effacing, turning, and abyssal echoing of traces. A somewhat similar encounter will occur in Cixous's work some fifteen years later, prompted by another work of mourning, that for the departed Jacques Derrida. In the central section of a book, *Insister of Jacques Derrida*, that pursues infractions of the rules of punctuation such as have marked all her work, Cixous is drawn, through another rediscovery of a text left in suspension or abeyance, to the point of a *point*, a literal, graphic dot. The series of parallels with the beginning of *First Days of the Year* is uncanny, beginning with the orality of writing, relations to one's maternal, and other tongues, reference to Celan, questions of victimization and scapegoating, and life after life.

Having agreed to write a paper for a conference in Barcelona in 2005 that Derrida had hoped to attend ("I will come perhaps naturally. Or else: I will come, perhaps, naturally"[23]), Cixous finds herself blocked by grief. She then stumbles upon a draft, handwritten manuscript that Derrida had mailed to her from Argentina back in 1995, an early version of his contribution to what would become their cowritten book, *Veils*. He had sent it to her in a sealed envelope and made her promise not to open it until, or because of, some unspecified eventuality that is also a theme of his text, an eventuality rendered as a "verdict."[24] She keeps that promise until now, when she opens the envelope and reads Derrida's manuscript:

> Look at it closely. It is to be seen, you see that: it is drawn at least as much as written, it is drawwritten, a breathless [*haletant*] self-portrait. One *sees the breath,* one sees the wind push the sail, rush the fabric. . . . I would write a book to de-pict or re-picture this painting with its quivering signs that look like they've been hurled from a brush, with its letters changed into the beating of wings, of lashes [*cils*], of see's minutely sown [*de si au semis minutieux*] *with living punctuations, of highly eloquent silences.*
>
> Look at this portrait from left to right, from top to bottom, it can also be read vertically like a poem—which it secretly is. One would have to be able . . . to remain faithful to the exact alignment [end of paragraph, no punctuation] (98, gloss, ellipsis, brackets, and emphasis added)

The title of Derrida's draft, "Points de vue piqués sur l'autre voile" (Points of view stitched on the other veil), to become the subtitle of "A Silkworm of One's Own," is replete with wordplay. *Points de vue* is heard as "points (of view)" but also as *point de vue,* "no view, none at all," hence no points at all. But the points are also "stitches," and they both prick or pierce by being sewn into the other veil (or sail) and, with slight manipulation of the syntax, are said to be "pinched" or stolen *(piquer)* from it. Those resonances echo through Derrida's writing and into Cixous's reading of it. All of my themes are present in her passage just quoted, in particular, breath/breathlessness of expression, punctuation that lives and speaks as silence, the spatial appearance of poetry. But as the photographic reproduction of Derrida's text shows (*Insister,* 113–120), what is most specific about the writing in purely graphic terms, what is most seen when one looks at it, is first the diagonal line with which, in most cases, he crosses out the entire page, second the arrows that propel the reader from page to page, and third his use of a point, precisely a •, to separate sections of the manuscript (Figure 3). The significance of such elements is already implied in the paragraphs from *Insister* just cited, but the encounter between Derrida and Cixous that takes place across the manuscript comes explicitly to be concentrated in the play of presence and absence, life

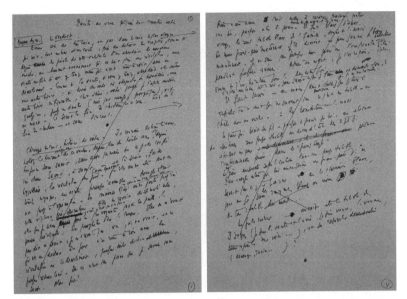

FIGURES 3. Derrida's "drawritten . . . breathless self-portrait." Photographic reproduction of Derrida's manuscript for "A Silkworm of One's Own," reproduced
in Cixous, *Insister* (Paris: Galilée, 2006). Reproduced with kind permission of
Hélène Cixous.

and after-life, represented by the dots. The *point* becomes the site of a
text-to-heart encounter similar to that she has via Celan's "Cello-Einsatz"—
"guessing from the rhythm [*tournure*, "turning"] of my breathing that I
was hearing him and listening to him" (102)—but which becomes more
complicated once Derrida's text not only speaks *to* her but at the same
time speaks *of* her, which is the case in "A Silkworm of One's Own" and
various other writings. Then it becomes a question of giving herself an
existence, a position, and a situation both inside and outside what she is
reading: "What kind of reading can one perform when one finds oneself
to some degree invited, lodged within the text to be read?" (107). Her
somewhat self-deprecating response will amount to seeing herself as one
of his worms (s. *ver*, pl. *vers*), which is also necessarily heard as one of his
verses (s. *vers*) or a line in his poem, or indeed, as a creature or life-form
whose drooling produces and weaves something of the fabric of the writing. But more specifically, her answer to the question of how to respond
to being lodged within the text, later reformulated twice as "where to

place myself?" will involve becoming one of those points, a black dot: "I find suddenly that I resemble somewhat that •, the black dot that floats between your stanzas in the manuscript. . . . This • would be altogether me" (110). As such, she accepts to be a problematic node within the textual system—a *point noir* is everything from a black stitch to a "blackhead" and a "traffic snarl-up" and could perhaps suggest a black hole—and even submits to being no point, none at all, disappeared, for when *Veils* gets published, the dots are no longer anywhere to be found:

> This • would be altogether me: A point of (non-) view. I mean a point without sight, a pupil without light. • around which, from which, you execute your fascinating dance of Veils. But this • goes and disappears in the book *Veils*.
>
> Where to place myself, now? (110)

Where now? That has in a sense been my question throughout this chapter: where now or how now? Where am I, or how am I, Cixous asks, once the *point* or *point not* that is "altogether me" goes and disappears? How do we speak of that empty space produced by the absence of a mark; how do we speak of the trace of such an absence? What form of existence, or, as I prefer, life, do we give to it? For "living punctuation [*ponctuations vivantes* in the plural]" is Cixous's precise term for the graphic, pictorial poeticization she sees in Derrida's manuscript. What form of life, then, would punctuation be? How does a point live, as long as it is a point, and how does it live once it disappears? That has been my question, a question concerning not just these aberrant or contrarian punctuation marks but every mark constituting a text to the extent that it necessarily involves such a play of tracing between presence and nonpresence.[25]

For this has never been about anything but life, from the tears, sobs, and heaving chest of Cixous's encounter with her lost book thirty years past in *First Days of the Year,* where the tipping of a teetering prose into poetry is a desire to "advance, by beginning at the end, death first, life next, next teetering life [*la vie chancelante*]" (*First Days*, 9), a chancy, secret, and *Celanating* life, to this continuing dialogue between a Derrida on the side of death and a Cixous on the side of life:

I am reading you. From-the-first-day you speak (to me) of dying
Book after text speech lecture. Since 1963. Already in 1955.
...
– But who is talking about living? (The tone? How to translate it? It's
you, it's he who is talking about living.) Living, I don't think about it. I live.
Living? you say. And from the first days he asks me if I believe it is
going to last.[26]

This will always have been about how to live on or in a point, a mark, a trait,
a trace that is both divisible and *disappearable*, about what sort of breath
one draws, is drawn, by means of that graphic eventuality. And furthermore, this will always have been, not simply a formal aesthetic analysis or
reflection, but an interrogation that goes to the ethical and political heart
of whether and how literature brings to life, how it inanimates the real;
how it transforms bodies into marks on a page by means of something
other than a simple move from living flesh to dead writing; and ultimately,
how it deals with suffering bodies, even bodies that have been disappeared
in the most intolerable or inhuman way—how it expresses the inexpress-
ible, dealing with the challenge of what cannot but must be written.

Cixous brings to the question of the living point the attention of a
physicist. In *Portrait of Jacques Derrida as a Young Jewish Saint*, her earlier
book centered around Derrida's "Circumfession," she has a chapter entitled
"Point Donor" ("Point donneur," which also reads as *point d'honneur*, a
matter of honor), which begins thus:

> The point to which the *point* is a thorn in his side, the whole corpus of
> his work bears the stigmata. The point is the absolute unity, without di-
> mension, says mathematical discourse ... [but] the fundamental axiom
> of everything he says everywhere is the *divisibility of the point*.
>
> Everything he writes, everything he thinks is a protest against the point
> as indivisible. He writes, divided, in order to divide it, the point. ... Never
> does he put a point, a dot, a period, the word, the sign, without a shudder.[27]

And shortly thereafter, having cited the opening lines from "A Silkworm of
One's Own"—"Before the verdict, my verdict, before, befalling me ... stop
writing [*ne point écrire*]"—she emphasizes that the use of the word *point*,

however much it reinforces negation in purely grammatical terms, functions as an opening and an enlivening in the very face of that negation:

> To write without a period, without end ... before it is too late, what an idea!
> He doesn't say *not to write* you will note, he points to the point, he makes his point by point. He didn't say not to write [*ne* pas *écrire*]. *Pas* means *step* and has to do with walking. *Point* has to do with space. . . .
> In the case in point, [the sentence] will have come along, twisted with apprehension, bowed by the wind of imminence. Imminence of what? It is always death that *immines,* one doesn't even need to name it, one must not. When there is imminence one has no other resource than to live on the point [*vivre sur le point*]. . . . One ought to give meticulous thought to the expression that recurs in *Un Ver à soie, on the point of.* To think the quasi-infinite multiplication of the point before it becomes final. . . . In this way one weaves toward the final stitch, or the last word: the smaller than any finite quantity and yet this point is not nothing [*ce point-là n'est pas nul*]. . . .
> Before death there is lots of time, just before death there is still time. An infinitesimal space but a space nonetheless.[28]

Cixous's situation, the way she finds or places herself within Derrida's draft for *Un Ver à soie,* could thus be interpreted as her way of giving a lesson—the pupil become teacher—on that type of living on a point, as a point. It is a lesson that Derrida himself has already given, at least in the sense of imparting the information. For example, in *H.C. pour la vie,* the emphasis of whose title should not escape us, he continues the contrast between the two sides on which they fall by emphasizing ultimately that "life has no other side ... there is only one side, the one of living life. . . . No more edge [*bord*] for me. No more death, maybe, since life has no opposite edge" (47–48). Still, it is as if she is daring him here to put that into practice, to climb down from his crest or razorback edge and learn to inhabit a point, to live it despite the danger and fear he acknowledges in "Circumfession" of "dying before the end of a long sentence, period," despite the risk, as Cixous characterizes it, of being "nothing but a fullstopendofsentence [*unpointc'esttout*], a writer who's nothing but a point," rather than the "sentence as long as the world" that he aspires to be.[29]

If, then, we are to look at Derrida's draft in the way that Cixous was inviting us to read it with her, to see it drawn as much as written, we would have first to see a design with Derrida in it, "a breathless self-portrait." But it would be breath*less* only to the extent that it is all breath, all breathing or puffing *(haletant)*, a breath that "one sees, one sees the wind push the sail." Looking again, though, we would see a poem, passing our eyes down from line to line to parse a Broodthaers-type pictorial series of lines or verses, uneven spurts of exhaled scriptural expression. Here and there he would be observed stopping to catch his breath, resting within one of the black dots he inserts, providing space, as Cixous sees it, for her as partial object of the writing and also reader: "This • would be altogether me." Nothing but a simple point, but a living one nevertheless. For, humble or self-abnegating as her self-description be, its *point* is not for all that nothing. From her point of view/nonview, she nevertheless waits for her chance (a few lines later, she will write of seeking a mouse hole), as does every object of seduction (and note that *he* is dancing around *her,* to the extent that the sexes are that simply differentiated). Similarly, if as pupil she is the object of (his) pedagogy, the roles might change as we just saw, making his point hers; and if, as pupil in the optical sense, she consents to being eclipsed ("pupil without light"), shrinking to a tiny iris that seems to fade entirely, it would precisely be in order to better live on the point, to live in and as the reflex automobility or physiological contraction that we have considered to be a type of origin of life—to live in that way the chance of teetering life, a tiny life or a very little death, it becomes difficult to distinguish between the two once they are experienced on the same side. In this way, she meets the picture of the poem of him beyond the heavy breathing, instead in the place where his breath "turns," finding herself there as •, as point of view or nonpoint that goes and disappears. Having become, therefore, the very "living punctuation," a "highly eloquent silence," she will be as eloquent in appearance as in disappearance, indeed perhaps all the more poetically *élocutoire* for having disappeared, as Mallarmé famously stated in his *Crise de vers.*[30]

Paul Celan's 1960 Georg Büchner Prize address, *The Meridian,* begins like something from Descartes's *Meditations,* all puppets, monkeys, and automatons. It weaves in and out of Büchner's "aesthetic conception," expressed by his character Lenz in the work of the same name, which requires that in art "what has been created ha[ve] life" rather than being variously "a puppet-like...childless being," something that appears "in the shape of a monkey," or "nothing but cardboard and watch-springs" (*Meridian*, 4, 2).[31] Now, on one hand, that hostility to a lifeless art is not entirely attributed to Büchner himself, for although "Lenz, that is Büchner, has...only disparaging words for 'Idealism' and its 'wooden puppets'" (4), Celan hears in Büchner's voice the sense that "for him art retains something of the uncanny" (5). He will therefore attempt in *The Meridian* to develop a very different aesthetic principle than that based on the simplistic choice between an art that is contrived and one that somehow manages to remain alive. On the other hand, according to a note among the materials that contributed to the composition of the address, Celan "confessed" to sharing that hostility, which goes some way toward explaining the ambiguities in his thinking that I shall shortly discuss.[32] As is well known, that thinking, as developed in *The Meridian*, precisely posits a distinction between an aesthetics mired in the question of lifelessness and a poetics that turns on the breath, the encounter, and the meridian.

Celan's poetics resists, on one level, the pneumatic expressivity that we have presumed to reside at the traditional base of poetic creation: "The poem is absolutely not, as many believe, the result of some kind of "expressive art / Ausdruckskunst" (153). By the same token, however, it remains intimately a question of breathing: "On breathroutes it comes, the poem, it is there, pneumatic" (108); "poetry is...breath-marbled language in time" (110); "the poem: the trace of our breath in language" (115). In order that such a poem come to constitute an encounter, therefore, some distancing will have to take place vis-à-vis that idiosyncratic, egological breath. Art will have to "creat[e] I-distance," such that "poetry, like art, moves with a self-forgotten I toward the uncanny and strange" (6). But the same distancing or uncanniness risks earning for poetry the reproach of "obscurity" (7). As a result, another operation is required, another form of distancing, one that will work, as it were, as a parallel companion to "the darkness attributed to poetry for the sake of an encounter," so as to produce "two strangenesses—close together, and in one and the same direction" (7). Paradoxically, therefore, a doubling of distancing is what manages to keep poetry on a conciliatory path. What prevents the poem from getting lost in its own distant darkness or dark distancing is precisely the experience of uncanny doubling otherness, an otherness that, for being close to the one, becomes all the more absolutely other. Only on that basis, "perhaps"—the word occurs ten times within a short passage of the address—will the Medusa's head shrivel and the automatons break down, enabling the poem to speak in the cause of, "on behalf of... who knows, perhaps on behalf *of a totally other.*... The poem wants to head toward some other, it needs this other" (8–9).

Now it might seem that Celan has redeemed the distance he had inserted into the poetic process in a classical manner, transcending, indeed, destroying, the artificial contrivances of art in ways that would satisfy Büchner's Lenz, allowing for what is created to return to life and perform an organic communication with another. But apart from the doubling of otherness just referred to, two other somewhat enigmatic terms are introduced at this point in the Meridian address to describe how a potentially contrived art effectively becomes poetry. The first is the "breathturn [*Atemwende*]": "Who knows, perhaps poetry travels this route—also the route of art—for the sake of such a breathturn?... Perhaps the poem is itself because of this" (7–8). The second is the date: "Perhaps one can say that each poem has its own '20th of January' inscribed in it? Perhaps what's new in the poems written today... is the clearest attempt to remain mindful of such dates" (8). As I shall shortly read them, those two elements make a poem distinct from a mechanical artifice, not by returning to any natural expressivity, not back to a form of life that would be untouched by anything uncanny [*unheimlich*] or unhomely, a life safely at home for never having traveled. Rather, the breathturn and the date allow the poem to voyage unendingly into the space of the encounter, an encounter that can become a "conversation... often a desperate conversation," letting "the most essential aspect of the other speak: its time" (9).[33] And if there is ultimately achieved "a kind of homecoming," it will nevertheless come about as the inconclusiveness of a "sending oneself ahead toward oneself, in search of oneself" (11). Finally, at the very end of the address, to account for that complex and paradoxical movement or journey, Celan will return to his title and, as it were, happen upon the word for such a poetic tracing of a path. There he introduces a third term to describe the "immaterial, yet terrestrial... circular," something that "connects and leads, like the poem, to an encounter," calling it a "meridian" (12). One might imagine him in the process of composition, musing and looking at his son's globe, perhaps twirling it, and deciding that, however extravagant the tropical digressions of poetry be, however contrived its conventions and contraventions, nothing better figures how it traces its path, as relation to the real world and as aspiring to a type of universal communicability, than the latitudinal gradations that both link and separate the inhabitants of the earth.[34]

The relation between the meridian and a date is obvious, as the former provides the basis for tracking the passage of sunlight across the earth's surface. Hence the International Date Line itself, opposite the Prime (Greenwich) Meridian, forms a most contrived and crooked example of it, charting its limping path through the Pacific Ocean in an attempt to impose the least inconvenience on the populations that stand astride it. Less obviously related to the meridian is the *Atemwende,* or breathturn, unless we are to think of it as the very turn toward, or into, tropological uncanniness, the I-distancing or pneumatic self-exile, a provisional respirational expiration that Celan understands as the condition of possibility of the poetic odyssey. In that respect, it would be akin to the broken heartbeat, the heart that beats as its own syncopation, such as we shall see in our discussion of Jean-Luc Nancy's "ex-scripted" love in chapter 7. But a similarly apposite reading would understand Celan's breathturn within the context of the coming to poetry, or *Celaning,* of Cixous's absence of punctuation that I have just discussed at length. For if the Medusa is slain and the automatons rendered inactive by the breathturn, it would be less because the pneumatic exhalation has resumed as though without interruption than because there has been activated through the turns, pauses, caesurae, spaces, and disappearing periods of Celan's and Cixous's writing something I am calling inorganic, automatic, or inanimated life.

As I have suggested, Celan's *pneumaticism* concentrates on a sense of rhythm that is as much about the interruption of breathing as it is about a traditional idea of poetic expressivity. In drafts and notes for the Meridian address, he repeats his mother's adage, "What's on the lung, put on the tongue," which makes the "breathroutes" that the poem takes pneumatic more in the physiological than the metaphysical sense, blown in pulmonary rhythm, precisely as cola that constitute "breath-units" rather than as exhalation of the heart or soul (51, 108). Similarly, the poem's syntactic force derives from what Celan calls a "speechgrille," understood as a "porous" system of "speechspaces" that emphasizes the interval and "stoppages of the breath" (109). "There is a 'poem in the poem,'" he notes. "It is in . . . each interval" (103). Thus, whereas one also finds reference to the breath as "inspiration," to the poem as having "the liveliness of mortal soul creatures" (113), there is little doubt by the time it comes to the "breathturn" that priority is being given by means of that word to the sense of a falling silent: "it faces you with silence . . . it takes your—false—breath away: you have come to the breathturn" (123).

Indeed, the word *Atemwende* that gets repeated and underlined in the final version of the address would seem to have been carefully chosen to signify a radical deflection, a diverting or detouring of the breath into silence, rather than the *Atempause* to which Celan also refers (8). When the word *Atemwende* is first used, it develops out of the context provided by the previous sentence, where Celan goes back to his prime example and primary reference in Büchner's *Danton's Death*. That example is Lucile Desmoulin's cry of "Long Live the King" shouted out at the scene of her husband's execution and in response to the Terror. Celan hears Lucile's cry as something other than a declaration of loyalty to the ancien régime, rather as a counterword (3) that precisely catches in the throat and stops the heart: "'Long live the king' is no longer a word, it is a terrifying falling silent, it takes away his—and our—breath and words" (7), or as another translation has it, "it robs him—and us—of breath and speech."[35] Indeed, the German verb *verschlagen* that is used there ("es verschlägt ihm—und auch uns—den Atem und das Wort") necessarily conveys via its root *schlagen* a certain violence. It strikes us dumb, makes our heart leap, and suddenly takes our breath away; it "cuts" the breath *(couper le souffle),* as the French would say. We should understand that interruption of the breath, its violent othering, to function as the originary strangeness, the very caesura or cut that constitutes poetry, the *poietic* or *physical* uncanniness that is in operation at the pneumatic origin itself, well before there emerges any artistic, artifactual, or artificial Medusa head or automaton. Without it there could be no passage along a meridian toward an encounter, no way at all for poetry.

In that sense, the poetry of a diverted breath would be both space and point, no more the punctuation of the empty space of breathing by sound and speech than the complex negotiation of breath and absence of breath. For neither would that breathing and turning from breathing be a matter of a simple alternation between exhalation and inhalation; rather, it would concern the tensions, extensions, gasps, and protractions through which the regularity of the breath is called into question, threatened by something within it that is foreign, wholly other to it, but that must be embraced by it, articulated within it. And in the same sense, what is read as poetry carried along by the *Atemwende* would be no more the filling of the blank space of the page by the points, lines, and letters of utterance than the intricate interruption of the written *trait* by a whiteness that functions as if on the other side of it, as the swirling interaction of spacing and visible accretions that produces legibility, producing that legibility as much, therefore, in the space itself as in the sedimented concentrations we call writing.

That would be how poetry comes to life, not simply as communicated breath of the muse, not simply as speech given life by metric regimentations, however creative, derived from respiration, but as the punctual interruption that inanimates through that very absence. For as I have tried to show, poetry not only reinforces but also disturbs the functioning of a punctuation derived from the pneumatic intonings and cadences of speech; by its diacritic interference within the relation between speech and writing, it not only comes to a point of emancipation vis-à-vis its supposed oral origins but also sets in train a wholly different functioning of what we might call negative versus positive materiality, a wholly different articulation of what constitutes the present and the visible when it comes to writing, a wholly nonnatural conception of what makes itself present and visible as life-form. If, then, we were to give a strong interpretation to Celan's reconfiguration, via the breathturn, of the opposition between a contrived and a living art, it would be by reading in it a sense of life that persists or survives in poetry because of, not in spite of—indeed, that is enabled by—*the very interruption of its naturality,* the fact of its being traversed by a radical otherness, an otherness whose structure, as we have consistently insisted, cannot not include the nonnatural, the prosthetic, the artificial, the inanimate.

As a token of that emphasis in *The Meridian*, one can note the references in preparatory and posthumous materials to Freud's *Beyond the Pleasure Principle*, specifically to the primordial "desire of life to return to the inorganic," and Celan's linking of Freud's formulation to "forms" as homage to the self-legitimacy (autonomy, entelechy) of the inorganic (*Meridian*, 165, 190). The logic expressed there cannot mean that form is an encroachment of the inorganic that implies the progressive decomposition of poetry, its death wish, for that would be contradictory to everything he has had us understand so far. Rather, it would be a matter of recognizing the inorganic or inanimating drive that operates through life from its origins to its most persistent and innovative forms. From inorganic to organic, there is indeed a passage across an opposition, a point or tiny animated speck appearing where previously there was nothing recognizable as such. A • where previously there was " ". But once there is such a passage across that opposition, it can no longer be the same opposition; a very different relation develops between the blank and the •, between the • and its other side, between it and the wholly other of its same.

Such a nonoppositional relation between life and nonlife, or between animate and inanimate life, is what lies along the path of the poem, in the space where the breath turns out of itself, not simply into a silence that signifies asphyxiation, but rather into a form of survival. It is Cixous's teetering *(chancelante)* but still so "fortunate" life *(si chance)*, a life concealed or clandestine *(si celante)* but still as eloquent as the music Celan makes heard playing from behind the pain in "Cello-Einsatz." It is also, however, a type of vulnerable, bare life. For in "Cello-Einsatz," along the way to the potential optimism of its final stanza, where everything is not only less but also more *(alles ist weniger, als es ist, alles ist mehr)*, there will have been a reference to the breath, but one that is haunted by clouds of fire: "two / smoke-clouds of breath [*zwei Brandwolken Atem*] / dig in the book / which the temple-din opened / something grows true / twelve times the / beyond hit by arrows lights up / the black- / blooded woman drinks / the black-blooded man's semen / all things are less than / they are / all are more."[36] In his evocative translation, Pierre Joris has "blaze-clouds of death" for "zwei Brandwolken Atem."[37] Poetry is, of course, no guarantee, and ultimately no protection, against immolation. In the space of the breathturn, it finds also that sort of danger.

Celan's 1967 collection *Atemwende* testifies consistently to ash and cinders. And indeed, as we know, the Shoah is a constant, perhaps necessary referent in all of his poems. In the same context, *The Meridian* is systematically referred to not just as Celan's major statement on poetics but as a, or the, major statement on poetics after Auschwitz. Yet the final draft of the speech, and indeed all the preliminary versions, avoid explicit references to the Holocaust. The closest factual or historical reference to it is given by means of the triple mention of a precise date, January 20, which has been identified as the date of the infamous "final solution" conference held at Wannsee in 1942.[38] In certain drafts, it is called "our 20th January" (65, 67), and in what the editors consider to be one of the earliest sketched outlines of the speech, he formulates it this way: "We write always still, today too, the 20th January, this 20th January, to which there has since been added the writing of so much iciness [*zu dem sich / seitdem / soviel Eiseges hinzugeschrieben hat*]" (68, translation modified). Yet when he introduces the date for the first time in the speech, it is within a citation from Büchner's *Lenz,* whose protagonist "on 20th January walked through the mountains" (7). In order that each poem have "its own '20th January'" (8), and in order that Celan can say "I had written myself from one '20th January,' from my '20th January'" (11)—those are the later mentions of it—the date will have had to be wrested from Büchner and recontextualized by means of a complex play of quotation marks. As a result, it can no longer refer only to the Wannsee conference and so to the Shoah. It seems quite clear that even presuming it was Celan's intention to have the individual or collective memory jogged and to have us realize that January 20 should be as instantly recognizable as November 11, June 6, August 6, or any other high-profile date, he was prepared to risk, indeed to insist, that the date would also migrate beyond the specific context of the Nazi final solution.

Now, two complicated literary questions have been opened up here. The first is that of the hermeneutic operations of literature in general, the question of whether, to put it simply, identifying a reference in, say, a poem—and we are compelled, I think, to read *The Meridian* as a type of poetic utterance—adds to or subtracts from the signifying operations of the text.[39] Celan's reference to Wannsee appears deliberately oblique, and, although myriad doubts should remain concerning what *deliberately* and *oblique* really mean in such a case, we have to ask whether we are doing his poetics a favor by identifying the date as that of the Wannsee conference. That is rendered all the more urgent by the second question, namely, the fact that dating is itself a complex phenomenon in Celan's poetry, as Derrida has analyzed in great detail in "Shibboleth," where he discusses the use of dates in a series of poems and devotes a number of pages to the January 20 of *The Meridian.* A date is for him a (non)singularity that is comparable to circumcision, on one hand, and the specific delineating code or "shibboleth," on the other:

> For example: there was a 20th of January. Such a date will have been able to be written, alone, unique, exempt [*soustraite*] from repetition. Yet this absolute property can also be transcribed, exported, deported, expropriated, reappropriated, repeated in its absolute singularity. Indeed this has to be if the date is to expose itself, to risk losing itself in a readability.[40]

Derrida draws attention to the series of "buts"—punctuating the discourse in alternating intermittence with the "perhaps" referred to earlier—that follow the central reference to January 20 in *The Meridian,* where "what's new in the poems written today" is said to be "the clearest attempt to remain mindful of such dates" (8). After asking, "But don't we all write ourselves from such dates?" Celan declares, "But the poem does speak!" (8), which means, for Derrida, that "despite the date, in spite of its memory rooted in the singularity of the event, the poem speaks.... If the poem recalls a date ... the date on which it writes or of which it writes ... yet it speaks! to all, to the other, to whoever does not share the experience or the knowledge of the singularity thus dated."[41] However much the poem owes itself to the date, as the "provocation" or inspiration that gives rise to it, once it assumes its own voice and speaks as itself, indeed, to be able to perform as poem rather than simply stating a fact, it will have to "release itself from the date without denying it, above all without disavowing it. It absolves itself of the date so that its utterance may resonate and clamor beyond a singularity that might otherwise remain undecipherable, mute, and immured in its date."[42]

Derrida's analysis in fact combines my two questions and shows how the date functions in the same way as any form of poetic encryption, any singularizing signatory effect that, however much it appears to rely on a particular reference—which we might or might not be able to identify and relate to the history or biography of the writer—necessarily offers itself for reidentification or reassignment beyond the specific memento of its original inscription. Hence, on one hand, Celan's reference names the Shoah, but that comes down not to whether, in *The Meridian,* he wanted that significance of January 20 to be instantly recognized, or, indeed, whether he wanted to send us on a search so that we might unearth other information along the way; rather, it comes down to the simple fact of the identification of the date however it occurs, and even if it weren't to occur, and even if Celan were to have had something completely different in mind (which he does in the sense that he begins by presenting it as Büchner's January 20). And, on the other hand, he gives January 20 to every reader, to recognize or not, to recognize and respect, to recognize but ignore, to keep silent about, to bury as a footnote to history, even to disrespect or profane—as well as to appropriate as a reference to whatever might have happened on *their* January 20. It is as if there were to be found in the marks constituting that date the same play of blankness that we have been examining throughout this chapter, a blankness that highlights the date, underscoring and *emboldening* it, engraving it upon our memory, as well as a blankness that is an abyss where silence falls into a vertiginous embrace with its various others: the cries and screams of tortured victims, the ultimate silence of their death, and the silence of what within all that remains unspeakable.

The breathturn represents a deexplicitation with respect to the Shoah that is similar to that occurring with January 20. Earlier notes for *The Meridian* include this: "Only when with your most own pain you'll have been with the crooked-nosed and yiddy and goitery dead of Auschwitz and Treblinka and elsewhere, will you also meet the eye and the almond. And then you stand with your fallen silent thinking in the pause which reminds you of your heart, and don't speak of it. And speak, later, of yourself. In this later, in there, remembered pauses, in the cola and the mora, your word speaks: the poem today—it is a breathturn" (127).[43] Following the logic of that early formulation of the *Atemwende,* we can understand that encountering the positive poetic potential of life—the eye and the almond, which would later be thematized in "Mandorla," from *Die*

Niemandrose (1963)—requires dealing first with the murderous extremes of anti-Semitism; it necessitates being with the dead of Auschwitz and Treblinka and elsewhere. But subsequent to those experiences—presuming they could reduce to something we could simply experience, live through, and move on from—there would be a thinking that falls silent, that pauses like a syncope of the heart, a refusal to speak. The poem as breathturn is to come from there, out of those pauses, in the relations of stress and pause ("cola and mora"). The note suggests again that Celan's sense of abstraction accepts and even requires an elision of the Shoah, falling silent on it, pausing from it. But such an elision is not for all that a disappearance; rather, it is the basis for a transformation of it into "your word," like "your," "our," or "my" January 20. In that respect, Celan would agree both with Adorno that "to write poetry after Auschwitz is barbaric . . . it has become impossible to write poetry today" and with Peter Szondi, that "after Auschwitz no poem is any longer possible except on the basis of Auschwitz," as well as with Werner Hamacher's extrapolation that the Auschwitz basis for poetry is an abyss wherein "the poem can still speak only because it exposes itself to the impossibility of its speaking": "For this reason, Auschwitz does not become for it a historically limited fact, the murder does not become an unproblematical object of speaking, but the objection of a question that concedes its muteness and thereby admits that it itself has been struck by that murder."[44]

Parallel, or conversely, to that—for I am not sure how to characterize accurately the differential relation involved here—we might also read the elision of the name of Auschwitz by means of the breathturn as the "triumph" of the automatic survival of inanimate life. Not that the ashes of those murdered there will ever be resurrected, not that their memory will ever be anything other than a memory, a vigil and a labor of vigilant mourning. Nor am I by any means suggesting that poetry comes back into its own by trading on the Holocaust or by profiting from its victims. I am referring rather to the paradox that Celan seems to allow, whereby that unutterable horror speaks through its silence as well as through its mention. And I am suggesting that if the memory of the victims of Auschwitz lives on thanks to the turn of the breath out of itself, it is because there exists a conception of inanimate life—call it memory, or the archive—that survives beyond natural life. Without it, there can be no commemoration of the victims of the Shoah, no more than there can be any commemoration of any living being at all. That very concept of artificial, nonnatural, inorganic, or inanimate life is what permits such a commemoration rather than any presumption concerning a natural, organic, animate life that would carry over uninterrupted into memory and commemoration (or into poetic expression, as Büchner's Lenz would have it).

The interruption of natural, organic, animate life is at work in that very life from the first breath or heartbeat, and it runs the risk of stalling or forestalling that life from its beginning. Such an originary inanimation installs interruption as a necessary structure of life, allowing as much for what we ordinarily call natural death as for the unannounced or violent termination of life, while at the same time instituting, as similarly necessary possibility, the substitution for life of an inanimate, inorganic form of continuance or *survivance*. If Auschwitz occurs as the indelible name of a massive and intolerably violent termination of life, it also represents the interruptive abyss for which no possible name is adequate. In articulating that impossibility, as well as a certain impossibility of poetic expression in general, Paul Celan's poetry passes by way of, and even constitutes an embrace of, life that is not governed solely by what might be called the humanistic or even animalistic pathos of mortality or immortality. It means less rejecting such a pathos—his poems are after all perhaps the supreme example of it—than entertaining a nonoppositional relation between life defined by autoaffective sentience, on one hand, and instances of contrived automatism, compulsive repetition, or cantilevered caesura such as also characterize those poems, on the other.

By means of Celan's detour along the breathpath that he calls the *Atemwende*, in that self-ellipsis of the breath and self-defeat of poetic expression, Auschwitz occurs via a type of nonmanifestation; it occurs as a blank space whose other side is the bottomless abyss into which there disappears each death *named* but each death *senseless* of those who died there, and—for the metonymic capacity of that blankless is limitless— those who died at Treblinka, indeed the dead and wounded of every other Nazi, but not always German, death-dealing machine from Janowska to Jasenovac and every other genocidal massacre, finally, however differentially, every end of life of every living thing past and yet to come.

If there is to be a poetry after Auschwitz, it will have to provide the

capaciousness for that immeasurable and imponderable prosthetic possibility; it will have to take that on as part of itself in the very place where
it sings and breathes, to carry that iron lung on its tongue, to bear that
imponderable load in the very place of its rhythmic transfer. Only then,
following both Büchner and Celan, or in both ways that Celan seems to
want, will what has been poetically created have life.

All of that falls, emerging as much upward out of it as disappearing into
it—interruptive, irruptive, eruptive—within the abyss of expressivity and
expressibility of black points and white spaces on the page. The direction
of it cannot be determined, no more than can the speed of it be mea-
sured. "Absolute speed, the speed that absolutely economizes on speed,"
is how Derrida characterized Cixous's homonymic substitution, which
I analyzed earlier as an abyssal enfolding of language comparable to the
play of punctuation and blank space.[45] But Derrida also relates speed to
a form of potency or potentiality that is akin to the force of life, tracing
it through Cixous's lexical and semantic networks ("*la vie, je vis, vitesse,
vision, vivacité, vivement,* etc."),[46] in particular through her idiosyncratic
usage of the subjunctive form—*puisse, puissiez*—of the verb *pouvoir,* "to
be able," whose noun form means "power." For Derrida, that originary
jussive subjunctive is like a "let there be" of a divine creative instant such
as I'll return to in the following chapter, a "let there be" realized in the
instant of being enjoined or invoked. The power, potency, or potential of
it "no longer designat[es] the virtuality, the potentiality, a *dynamis* that
one could traditionally continue to oppose to *energeia*. No, what arrives
according to this mighty power of the "might," of the "would that it, he,
or she might," really actually arrives, in real life. *It is life for life.*"[47]

My aim here has been to analyze the means by which, in the punctuation and blank space of a poem, the breath of literary expression crosses
over into inanimate life, and conversely, how what we understand as
historical fact and life, either microcosmic or macrocosmic, private or
global, survives in literature according to a complicated logic that is irreducible to an opposition between manifest and latent, living beings in
one place and simply words in the other. In both cases, in both directions,
in emergence or disappearance, life happens—is inanimated—across the
plane of the text. Perhaps one could say that life is exchanged for life, in

differential time-space. That would be especially so in the case of Cixous's "living punctuations," where there is a strange unuttered homonymy on one side and the other of a black dot, a form of silence that both appears and disappears and continues to signify. In such a case, utterance is suspended elliptically while imposing itself graphically (as . or •); or a semi-silent or silent inhalation, even an ultimate expiration, is seen resounding forcefully and proliferating exponentially all over the page (as " "). A dark dot disappears into its own black hole and/or a universe expands from nothing, exploding into the thin air of blank textual space: a poietic force of originarily cosmic inanimate life, life for life really actually arriving in the real life of the page as the instantaneous substitution of every end of line for its middle and every empty white "point" in between.

5

NAMING THE MECHANICAL ANGEL

Benjamin

Through Cixous, I have just concluded, punctuation comes to life as a strange unuttered homonymy on one side and the other of a black dot: not something emerging from a latency or silence that would have always harbored it, and not just a problem concerning the different forms of utterance that are graphic traces on one side and enunciated sounds on the other. Rather, the strange homonymy I was trying to describe constitutes the very relation of differentiated sameness between mark and absence of mark, between one side and the other of a single mark, or between the written sign and its oral other; even more specifically, it constitutes the absolute speed by which one substitutes for the other, the absolute speed of "the relation to itself as relation to the other."[1]

This chapter examines that homonymic speed in relation to translation: what happens when the absolute speed of a homonymic doubling encounters the absolute resistance of its untranslatability. Cixous, whose French is replete with homonymic wordplay, represents such a challenge for translation: "homonymy is . . . the crux [*croix,* also the "cross" the translator must bear] of translation: it is what, in a language, signals and signs the untranslatable. . . . The entire work of Hélène Cixous is literally, and for this reason, untranslatable, therefore not far from being unreadable."[2] But as Derrida points out, that untranslatability on the threshold of unreadability reinvests language with what is often considered to be its ancient and primary function, namely, naming: "When a phoneme becomes untranslatable, it begins to resemble a proper noun [*nom*]."[3] A proper noun and a proper name, the French is homonymic and therefore heterosemantic. But in the beginning, in the very beginning, noun and name were indeed indistinguishable. That scene develops as the subject of Walter Benjamin's essay on language in general and human language in particular, "On Language as Such and on the Language of Man."

"On Language as Such" obeys a type of progressively palimpsestic logic, and its shifting emphasis and successive conclusions help perhaps to explain why it remained unpublished in his lifetime.[4] In it Benjamin begins by positing—but "in no way metaphorical[ly]," that "we cannot imagine anything that does not communicate its mental nature" (*Writings 1*, 62). That means that everything, "in either animate or inanimate nature . . . communicate[s] its mental contents" (62), and the example he will turn to a couple of paragraphs later is a lamp that we presume to be very close at hand. Benjamin's position on language appears here to vary considerably from what would become the dominant view following Saussure, whose *Course in General Linguistics,* from classes in 1907–11, was published the same year (1916) that Benjamin was writing "On Language as Such." Even though he understands that the "mental nature" or "the mental being of the lamp . . . is by no means the lamp itself" (*Writings 1*, 63), he does not presume Saussure's arbitrary relation of "acoustic image" to "mental concept," and his view of linguistic operations seems closer to a type of mystical expressivity of the world. However, he immediately goes on to emphasize that whereas objects express their essence or nature, any expression that takes place via language necessarily involves communicating the essence of language along with any communicability of the object: "Language therefore communicates the particular linguistic being of things, but their mental being only insofar as this is directly included in their linguistic being, insofar as it is capable of being communicated. . . . The language of this lamp, for example, communicates not the lamp . . . but the language-lamp, the lamp in communication, the lamp in expression" (63). As a consequence of that idea of linguistic communicability, understood to take place less *through* than *in* language, language is always involved in communicating itself.

When it comes to the human, the principle that the "linguistic being of things is their language" implies necessarily that "the linguistic being of man is his language," but because the human speaks words, her mental being is communicated by means of the process of naming: "*It is therefore the linguistic being of man to name things*" (*Writings 1*, 64). Words are at base names, and language is in essence a nominalization.

If a lamp communicates not the lamp but the mental being or entity of a language-lamp or lamp-in-expression, the name communicates not the name but the mental entity of a language-name or name-in-expression that is, in turn, indistinguishable from language itself: "the name is that *through* which, and *in* which, language itself communicates itself absolutely" (65). Hence, when man names, he is vouching "for the fact *that language as such* is the mental being of man" (65). In other words, man communicates his mental being as language-man by means of naming, which in turn communicates its mental being as language-name, which amounts to expressing "language itself in its absolute wholeness" (65).

However, for Benjamin, this naming by the human is not performed as the communication of factual information from one human being to another, something he dismisses as the bourgeois conception of language.[5] A lamp may communicate itself as information to the human, but the language-man who names is ultimately communing with the divine: "*in the name, the mental being of man communicates itself to God*" (65). As Benjamin recasts his argument in the second half of his article, the importance of such an assertion will come to be understood less as reimposing a hierarchical chain of being—however much that appears as an inevitable consequence—than as reinvesting language, and speaking, with poietic force. If man communicates his mental being to God by naming, it is because in that way he participates in bringing the created world into being as he had by naming the animals in Genesis. He is not only in conversation with God but imitating his creative force:

> Man is the namer; by this we recognize that through him pure language speaks. All nature, insofar as it communicates itself, communicates itself in language, and so finally in man. Hence, he is the lord of nature and can give names to things. . . . God's creation is completed when things receive their names from man. (*Writings 1*, 65)

Benjamin's interpretation of discrepancies between the two creation myths in Genesis 1 and 2 concentrates on how God's rhythmic poietic riff of producing ("Let there be"), appreciating ("he saw that it was good"), and naming was interrupted when it came to producing man. As he reads Genesis 2, man arrives on the scene not simply as God's sixth-day repetition of the same refrain but as a being that remains nameless and instead has the task of naming conferred upon him. It is only after man has named all the other animals that he will celebrate in Genesis 2:23 by indulging in his own rhythmic and semantic élan and, as if in the same breath, name himself as man *(ish)* and woman as woman *(isha)*. By that time, man is not only channeling God's creative power in general but, indeed, is refining it by inventing the proper name, defined as "the communion of man with the *creative* word of God" (*Writings 1,* 69). Thus, when the Fall comes to pass—particularly the fall of Babel—it will be marked precisely by man's abandonment of that naming function in favor of the "externally communicating word"; language will then be reduced to a means, "a mere sign," which results in the plurality of languages (71).

Now, as we began by noting, Derrida identifies untranslatability with the proper name, yet in "Des tours de Babel," he also recognizes in that naming function the very condition of possibility of language. The proper name remains in one sense outside language but is in another sense the founding performance of language: "a proper name as such remains forever untranslatable, a fact that may lead one to conclude that it does not strictly belong, by the same right as other words, to language. . . . It does not belong to it, *although and because* its call makes language possible (what would a language be without the possibility of calling by a proper name?)."[6] Thus, whereas we would ordinarily associate translation with the Fall into plurilingualism marked by the Babel story, it is telling that Benjamin introduces the word or concept precisely in relation to God's conferral of the task of naming upon the human. For man has in effect already usurped divine privilege—thanks to God's invitation or command—as soon as he begins naming: he has already introduced the founding possibility of language and is already *in translation,* or at least *in untranslatability.*

As Benjamin reads it, therefore, an originary translation takes place between God's spontaneous creation (He speaks, and things come into being) and man's "receptive" creation (he speaks, and things and animals are named): whereas "the human word is the name of things," the thing in itself "has no word, being created from God's word" (69). Translation is the process by which the purely active spontaneity of a divine creation passes into the *passactivity* of what Benjamin calls "conception [*Empfängnis*]." By means of it, the uterine receptivity or welcoming within which the vestige of divine breath presumes still to operate assumes the sense of fecundation that is intrinsic to man's naming function. Translation is thus the transformative combination of God's spontaneity and man's conception, effectively the metamorphosis of autochthonous production into sexual reproduction: "For conception and spontaneity together . . . language has its own word . . . translation [*Übersetzung*]" (69). Such a translation, from ex nihilo one-way creation into a human production that depends on a form of relay or communication from God, institutes the first translation as a translation *into* language. In that sense, translation is what gives rise to language, at least to human language, before having anything to do with the transformation of one language into another.

It should be clearly understood that "translation" here goes beyond a simple figurative coinage, on Benjamin's part, of a word that will stand in for such other terms as "transformation" or "metamorphosis"; nor is it a term whose sense is to be limited to purely linguistic operations. It may well be a word that belongs to language, but if it is to refer to what operates before human language to give rise to human language, it will be rather *a wording before the word, a namelessness before the name, and a conceiving before the concept.* The passage just quoted reads in full as follows: "For conception and spontaneity together, which are found in this unique union only in the linguistic realm, language has its own word, and this word applies also to that conception of the nameless in the name. It is the translation of the language of things into that of man" (69). In my view, the strange syntax that Benjamin resorts to here should be taken, at least in the first instance, at face value. He does not write that "language has its word and *that word is 'translation'*"; instead, "language has its own word, and this word applies also to that conception of the nameless in the name. *It is the translation* of the language of things into the language of man" (69, emphasis added; Für Empfängnis und Spontaneität zugleich, wie sie sich in dieser Einzigartigkeit der Bindung nur im sprachlichen Bereich finden, hat aber die Sprache ihr eigenes Wort, und dieses Wort gilt auch von jener Empfängnis des Namenlosen im Namen. *Es ist die Übersetzung der Sprache der Dinge in die des Menschen*).[7] Of two syntactical possibilities, one is tautological—"the word 'translation' is the translation of the language of things into the language of man"—and the other densely elliptical: "the word for conception combined with spontaneity, and for the conception of the nameless in the name, is 'translation,' as in the translation of the language of things into the language of man." We would have to say that the word *translation* is not introduced as a word at all but as a conceiving mechanism on the basis of which words come to be; it is a process that is transformative or creative enough to allow a name to preserve within it a crypt or *chora* of namelessness and, conversely, to allow that namelessness to conceive naming, as if immaculately.

As Carol Jacobs has noted, the question of translation remains in Benjamin's work from beginning to end: "Perhaps no text of Benjamin fails to pose that question and to perform it."[8] Thus, five years before "The Task of the Translator," we see him developing a concept of translation that, as he insists in "On Language and Such," "it is necessary to found . . . at the deepest level of linguistic theory, for it is much too far-reaching and powerful to be treated in any way as an afterthought. . . . Translation is removal from one language into another through a continuum of transformations" (*Writings 1*, 69–70). But the first "step" along that continuum, what produces or structures it, is the transformation by which spontaneity cross-fertilizes with conception to found life as language. If we take Benjamin at his word, translation would, at the very least, be human life coming to be as language. We should understand the translation that he wants to find, or found, as a concept at the deepest level of the theory of language (*in der tiefsten Schicht der Sprachtheorie*; *Gesammelte Schriften,* II, 1, 151) to be the very concept itself of language, its founding concept as indistinguishable from the concept of founding. There where human language comes to be, there is translation as that very coming to be. Translation is thus the life of language, but no longer life derived from an origin, not the life breathed spontaneously into language by God; instead, it would be the originary technologization of that life as it is transformed from spontaneity to conceptionality. If God is, or gives rise to, the automatism of autochtonously created life, having it issue from his breath or sprout from the ground, then human language is the communicative relay—the "post" in Derrida's terms—by which, or *at* which, that life comes to be by becoming technologized.

Now it might seem that I am extracting a subtle nuance to convert or pervert what is in fact a rather traditional prioritizing of the divine spontaneous creation of the living world over man's derivative creation of artificial language—a classically linear trajectory from the origin to its secondary products, from God to man to language. But my insistence is based on two observations: first, Benjamin indicates that he is reading the creation not as narrative but as thematics or symbolics relating to the nature of language, such that the act of creation is itself interpreted as linguistic in the profoundest sense. If creation is achieved, completed by naming, that linguistic intervention is nevertheless a supplement that supplants by emerging as the structure of creation itself: "Language is therefore both creative and the finished creation" (68). Second—only superficially contradictory to what I have just noted—Benjamin's whole discussion takes place under the sign of the nonmetaphoricity I mentioned earlier: "There is no event or thing in either animate or inanimate nature that does not in some way partake of language, for it is intrinsic to each to impart its content [*denn es ist jedem wesentlich, seinen geistigen Inhalt mitzuteilen*]. This use of the word 'language' is in no way metaphorical [*Eine Metapher aber ist das Wort "Sprache" in solchem Gebrauche durchaus nicht*]" (*Writings 1*, 62, translation modified; *Gesammelte Schriften*, II, 141). Language is the literal word for how every event or thing, animate or inanimate, communicates what it is or contains. There is therefore no mute communication that finds its way into language but rather linguistic utterance that comes to be constituted as communication. The "emergence" of language through creation and naming that subsequently gets traced as the thematics of translation that we have just analyzed remains, logically speaking, under the sign of that literality: just as communication is literally language "before" it comes to be communication, so language is literally translation "before" it comes to be language.

In my view, it would therefore be incorrect to presume that Benjamin is describing a process whereby the human picks up where God leaves off through fatigue or boredom and gets done (for Him) the job of creation. Rather, the human intervenes in the process by translating it into something else, and that something else—the human form of creation or invention—is what we have come to call technology. As we saw in chapter 1, that technologization may be understood to "begin" with the mnemotechnological archive of the proper name. As man names each species in the Genesis narrative, he is effectively producing the afterlife of that species by giving it the word for itself, the name for it that will survive it, in the same way that he will later baptize his own progeny: he is giving them that form of lifedeath by means of translation, what we might call a *translative* name.

Nonmetaphoricity in relation to life, against a background of translation, is again foregrounded in "The Task of the Translator":

> It is evident that no translation, however good it may be, can have any significance as regards the original. Nonetheless, it does stand in the closest relationship to the original by virtue of the original's translatability; in fact, this connection is all the closer since it is no longer of importance to the original. We may call this connection a natural one, or, more specifically, a vital one [*ein Zusammenhang des Lebens*]. Just as the manifestations of life are intimately connected with the phenomenon of life without being of importance to it, a translation issues from the original—not so much from its life as from its afterlife [*Überleben*]. (*Writings 1,* 254; *Gesammelte Schriften,* IV, 10)

In a gesture that is reminiscent of his earlier essay, Benjamin here invokes life as the model for translation. The binding relation between a translation and an original that has no need or knowledge of it constitutes a fact or form of life, a connection constituted by life *(ein Zusammenhang des Lebens).* Translation is to the original what afterlife manifestations of life are to life; it is intimately and vitally connected even if the original appears to have no need of it. Far from intending translation to be understood as a lifeless and useless addition to a serenely impervious original, Benjamin wants conversely to argue for an enhanced status for translations, concluding that "in them the life of the originals attains its latest, continually renewed, and most complete unfolding" (*Writings 1,* 255). But in making that argument, in maintaining that a translation is related to an original in the way that an afterlife is bound up with the life from which it is presumed to be distinct, he performs an extrapolation that amounts to calling into question any pure organicity of life. And that extrapolation is far from inadvertent; it is an extended and explicit one, contended and maintained once again on the basis of nonmetaphoricity:

For a translation comes later than the original, and since the important works of world literature never find their chosen translators at the time of their origin, their translation marks their stage of continued life [*Fortleben*]. *The idea of life and afterlife in works of art should be regarded with an entirely unmetaphorical objectivity* [*in völlig unmetaphorischer Sachlichkeit*]. *Even in times of narrowly prejudiced thought, there was an inkling that life was not limited to organic corporeality.* . . . *The concept of life is given its due only if everything that has a history of its own, and is not merely the setting for history, is credited with life.* In the final analysis, the range of life must be determined by the standpoint of history rather than that of nature. (*Writings 1*, 254–55, emphasis added; *Gesammelte Schriften*, IV, 10–11)

Again, perhaps even more radically, translation here gives rise to life. On the basis of translation, Benjamin is arguing, starting from it, life will have to be redefined. If there is a coherent or binding life-relation between original and translation, then life cannot be limited to organic corporeality, then everything that has a history of its own is to be credited with life. None of that is metaphorical, he insists. There is, literally, inorganic, artificial, inanimate life. Just as, in "On Language as Such," we were to understand language in no way metaphorically as the means by which nature, both animate and inanimate, imparts its content, just as we were to understand that that language comes to be as translation, so now we should regard with an entirely unmetaphorical objectivity a relation of life to afterlife that is similarly determined by translation. This is the logic Benjamin appears to want to insist on: translation is what allows inorganic life to exist in its unmetaphorical objectivity; because there is translation, in its most literal sense of a transfer or carrying across from one language to another, a transfer that, as we know, is both possible— translation happens—and impossible as intact preservation of the original; because therefore, in the most literal sense, there is no such literal transfer; because, rather, there is the impossibility of such an uninterrupted passage, there is life. Only if we can find a way to make that logic work, will we be able to account for how an afterlife retains its vital relation to life; only once we have rejected the simple logic of organic causality will we begin to develop an unprejudiced concept of life. Life conceived of as translation constitutes a type of impossibility, or refusal of its own terminal necessity, that by which we would be able rigorously to separate it from an afterlife. Though we define it as finite, indeed, as the very definition of finitude, we will have to understand it as the translatability by which that finitude is everywhere transgressed or overcome. That, after all, is how life came to be, by somehow refusing to recognize the limitations of its own inertness, its own inanimation, or else by recognizing inanimation— inanimateness on one hand and *animatability* on the other—as its very condition of possibility, that is, by translating itself.

Benjamin's insistence on the nonmetaphoricity of a life that does not reduce to organicity is related explicitly to history: everything that has a history of its own, not what can be extended "under the feeble scepter of the soul" (*Writings 1,* 255), is to be credited with life. Afterlife is first *Überleben,* or *meta*-vital, before being the much less common *Fortleben,* an extension of life into its own archival survival. If we were to extend Benjamin's concept or life of history toward the other end of his short career, into his 1940 essay "On the Concept of History," and the corresponding "Convolute N" of *The Arcades Project,* we would be obliged to reconcile it with the materialist emphasis of that essay and its dissociation from history as a "process of transmission."[9] Such a reconciliation would mean, in the first instance, underlining another aspect of Benjamin's dynamic sense of relation—between spontaneity and conception, between life and afterlife—that we are taking to be defined and betokened by translation. That aspect is discontinuity, giving us to understand that the relatedness he wants to preserve across what seems at first glance to be an oppositional rupture does not reduce to an undifferentiated extension; on the contrary, translation is the very originary connection-in-rupture that gives rise to human language as spontaneity become conception (in "On Language as Such") and to the unmetaphorical objectivity of inanimated life (in "The Task of the Translator"). Translation continues life as discontinuous, allowing it to be denatured into history, denaturalized into what can be historicized: "In order for a part of the past to be touched by the present instant [*Aktualität*], there must be no continuity between them."[10] In the second instance, life as translation flashes up in a monadic moment that the historical materialist is presumed to seize. The interruption of the transmissive historical continuum is an explicitly appropriative gesture: "The past can be *seized* only as an image that flashes up in the moment of its recognizability. . . . Articulating the past historically . . . means *appropriating* a memory as it flashes up in a moment of danger."[11] Benjamin's examples in "On the Concept of History" are a Robespierre in Thesis XIV, who actively "blasts [*heraussprengte*]" a "now-time [*Jetztzeit*]" out of the continuum of history, and a virile historical materialist in Thesis XVI, who "leaves it to others to be drained by the whore called 'Once upon a Time' in historicism's bordello. He remains in control [*seiner Kräfte Herr*] of his powers—man enough [*Manns genug*] to blast open the continuum of history."[12] In a less testoteronal moment in Thesis XVII, he emphasizes the "constructive principle" on which materialist historiography is based, describing it as blasting a "specific life out of the era, a specific work out of the lifework. As a result of this method, the lifework is both preserved and sublated *in* the work, the era *in* the lifework"[13] (*Gesammelte Schriften*, I, 703). In the light of what has just been discussed, we would again have to recognize here the gesture of translation. If it might be exaggerating to read the historical materialist, who keeps himself—pure of historicist whoring—for consequential blasting, as the agent of a translation from spontaneity to conception, the references to another life that is constructed out of an era or a lifework do indeed return us to terms that are familiar: a sublative (*aufbewarht . . . und aufgehoben*) (703) irruption or interruption produces a constructed life. Thanks to that intervention, which functions as a visual (an image flashes up) and a nominal or conceptual recognition (the dialectical materialist takes cognizance of it), an era or a lifework gives rise to another form of life that is produced by contrivance; an inorganic or inanimate (after)life is produced that is not for all that any less life.

The dialectical materialist therefore bears some resemblance to man who names in Genesis: where God summoned the created animal world to present itself for man to name, history presents itself in flashes for the dialectical materialist's critical vigilance to recognize. Benjamin repeatedly calls those flashes images—"image is that wherein what has been comes together in a flash with the now to form a new constellation"[14]—which brings the task of translation back to a type of enlivening reading of the image. And, in a particular image that was very dear to Benjamin, indeed, an image that he acquired during the time he was composing "The Task of the Translator," the dialectical figure of history is less man than angel. I refer, of course, to the painting by Paul Klee that Benjamin acquired in spring 1921, *Angelus Novus* (Plate 1). The image and name of *Angelus Novus* was to have served as emblem and title for the journal Benjamin planned, but failed, to launch in 1921. The journal project preoccupied him for much of the first half of the 1920s, and the painting would continue to inspire him for the next and final twenty years of his life, receiving its crowning iconic reference in "On the Concept of History":

> There is a picture by Klee called *Angelus Novus*. It shows an angel who seems about to move away from something he stares at. His eyes are wide, his mouth is open, his wings are spread. This is how the angel of history must look. His face is turned toward the past. Where a chain of events appears before us, he sees one single catastrophe, which keeps piling wreckage upon wreckage and hurls it at his feet. The angel would like to stay, awaken the dead, and make whole what has been smashed. But a storm is blowing from Paradise and has got caught in his wings; it is so strong that the angel can no longer close them. This storm drives him irresistibly into the future, to which his back is turned, where the pile of debris before him grows toward the sky. What we call progress is *this* storm.[15]

We should try to understand the complexity of the operation Benjamin has set in play here. He takes the image and name that Klee gives the angel and invents them anew. An image is given by Klee, who names it at the same time "angel" and invention, innovation, newness. But it is as if the angel remains trapped and immobilized in its newness until Benjamin comes to rename or translate it as the angel of history. That translation is what will give it life, transforming it into a dynamic hermeneutic force, at once messenger and diviner, both blown by irresistible forces against its will, aghast witness to the devastation it retreats from, and flashpoint of recognition wherein indiscriminate debris becomes dialectical nowness.

In coming to this image that is presented to him, in order to rename its newness nowness, Benjamin has not only performed the inanimating function of translation but also radically reoriented the space of signification. For the angel that he describes, back turned to the future, face turned toward the past, situates wreckage, Benjamin, and any other spectator squarely in the space outside and in front of the image. We are to understand the space that we would normally regard as the background of Klee's painting, the space behind the angel, to operate as its foreground, its future, and the space in front of the angel, at its feet and beyond toward the spectator, as the past. Once the angel is said to be moving backward into the future as Benjamin sees it, then the perspectival organization of the image has to be reversed. We have to look at it as if from behind. Otherwise, the spectatorial space of the painting—what is outside it in front of it, where we stand, coming along later to look at the angel, face-to-face with it as its future extension—would be its past.

To repeat: according to Benjamin, the space inhabited by whomever the angel is looking at, for example, the spectator, is that of the *past*. The *future* for and of this painting is behind, on its other side. But according to temporal rationality, the future of the painting is determined by a spectator who comes along after it has been produced and hung on the wall, a future spectator, and according to spatial rationality, that spectator has to stand in such a position as to face the painting and look at the front, not the back, of the angel.

It is something of the same paradox that Benjamin introduces by having us understand a translation, something that comes to face an already existing original, something produced in the readerly and, as it were, spectatorial space of a work of literature, as its *after*life. Not just that it comes after, as does any future *survivance,* but that its distancing *(Fortleben)* is necessarily a form of rupture. It disrupts and contorts by means of a radical perspectival reversal; it requires the opening of dorsal space, whose materiality I prefer to imagine, rather than history, as a dynamically inanimating technology.

To the extent that it is corporeal, as in Klee's painting, and to the extent that we presume such a spirit to be natural, the angel articulates the encounter of nature with history, the transformation of nature into history that is Benjamin's prerequisite for life: "The concept of life is given its due only if everything that has a history of its own . . . is credited with life. In the final analysis, the range of life must be determined by the standpoint of history rather than that of nature" (*Writings 1,* 255). The appearance or incarnation of the angel thus represents an interference in nature,

nature's disjunction or becoming-other. It is something very different from a creation, and the disjunction out of nature is inevitable or has already occurred, I would argue, once the angel possesses a body, or a human face. That once the angel is visible it can no longer be simply an angel is something we have learned from the most famous case, the annunciating angel Gabriel, whose very presence, not to mention his words, represents something of a fecundation, if not a violation.[16] But that aside, such disjunction is certainly explicit once the angel is a figure drawn on the painted surface, once it is clearly no longer some ineffable spiritual emanation, some spirit or sprite. An angel embodied in ink, chalk, and brown wash is an angel fallen not just into incarnation but into technology; it has become artifact. The future it recedes into is a technological future that nevertheless occurs behind and in its back because it was always there, originarily, from the beginning. At the outside, one can say that the angel never was without technology, that a nontechnological angel has never been seen. Where anything resembling human corporeality begins, there technology begins also. If the angel doesn't see that, it is because it *is* that. In becoming history, in blasting itself dialectically out of the continuum, it has become the point at which, the flashpoint by means of which, as we see in the image, some technoanthropic form comes to be. Indeed, it does not look like what we would normally call an angel; rather, a bodily form emerges from, and defines itself against, a nebulous background, simultaneously revealing itself as a rudimentary mechanicity—intersecting lines, triangles, tubes, and scrolls—a prosthesis of proteiform organicity and marionette, animation and automation.

Having opened life to what has a history of its own, beyond organic corporeality, "The Task of the Translator" subsequently resorts to a traditional hierarchy of greater and lesser life, seeming to reinstall the structural priority of *physis* over *technē* that Benjamin has just called into question. The lesser form of life he refers to, in comparison to translation, is criticism. Benjamin, who produced a certain amount of critical commentary himself, refers to the Romantics, who, "more than any others, were gifted with an insight into the life of literary works—an insight for which translation provides the highest testimony"; however, they "hardly recognized translation in this sense" and instead "devoted their entire attention to criticism—another, if lesser factor in the continued life of literary works" (*Writings 1*, 258). Such a demotion of the task of criticism is in stark contrast to the theory of commentary that he develops, de facto but also in certain explicit terms, in *The Arcades Project*, and which he connects intimately to the project of the dialectical historian. Indeed, Carol Jacobs will read this reference to criticism as evidence "that when Benjamin speaks of 'translation,' he does not mean translation, for it has never ceased to acquire other, foreign meanings. One is tempted to read 'translation' as a metaphor for criticism."[17]

For my purposes, there is no meaningful distinction between criticism and commentary, and both imply translation: "every interpretation is a translation," says Derrida (*B&S I*, 336). Both are forms of exegetical explanation or interpretation that, falling within the same structural possibility as translation, cause a work to *live on otherwise*. Benjamin's theory is posited precisely on such a structural possibility, for it relies on a principle of interruption, on one hand, and montage, on the other. The recognition of an image as dialectical flash out of the historical continuum implies, by extension, a form of commentary that consists of combinations of such images, repeatedly described as montage. The theory of *The Arcades Project* is held to be "intimately related to that of montage," its method "literary montage," and its first stage "will be to carry over the principle of montage into history."[18] Commentary will therefore amount to juxtaposing images according to a dialectical principle of discontinuity. But montage seems to refer less to the cinematic theory of an Eisenstein than to the chance and incongruous combinations of surrealism (in the now of recognizability, "things put on their true—surrealist—face"[19]). For the effect of commentary appears to derive less from what the combination produces than from the very juxtaposition: "I needn't *say* anything. Merely show."[20] One might want to argue, as a consequence, that commentary is precisely distilled in Benjamin's project to something that approaches, and even strives for, its own inexistence: extract and combine, add nothing.

However, the "blasting" gesture necessarily means opening an interruptive or irruptive space. Commentary, criticism, exegesis, interpretation, and translation all presume the migration of marginal space into the textual interior. Dialectical intervention is necessarily disruptive in that sense, rending textual integrity. It means opening the very space of commentary, extracting the flash from the continuum so as to "insert" discontinuity, making that intervention itself at least an empty commentary, a comment involving emptiness, about emptiness, and at most an insertion that expands or proliferates without limit, for once there is interruption, no law is capable of controlling the extent of that rupture.

Benjamin will ultimately recognize the simplest operation of that principle where it might be understood to begin, in citation: "To write history thus means to cite history." For as he immediately explains, commentary ("writing history") derives from citation precisely thanks to the logic of *decontextualizing* just developed: "It belongs to the concept of citation, however, that the historical object in each case is torn from its context."[21] Citation presumes nonintegrity of context; it presumes extraction and concomitant disruption. Once one is in the business of citing—and the task of the commentator or critic is, in the first instance, that of citing—one has unleashed an iterability that allows for illimitable decontextualization and recontextualization, a textual fraying and a tearing with no hope of repair, but on the basis of which montage is made possible. That is the sense of Benjamin's "showing" without "saying": it means extracting and combining in such a way either that the juxtaposition itself produces its own comment(ary), such as the spark that the surrealists vaunted, or that the different elements juxtaposed in the montage comment on each other by virtue of their proximity one to the other, producing a new syntagmatic combination that is not for all that a new continuum, for it retains the ruptures on the basis of which it was first constituted. Hence, in the final analysis, not only will there be the showing without saying of citation without commentary, but even the showing itself is to be minimized, for "this work has to develop to the highest degree the art of citing without quotation marks."[22] In Benjamin's citing without the marks of citing, however, one cannot expect to find montage as continuity editing, a seamless renaturalization or reimposition of the continuum, for that would be to negate the "break with vulgar historical naturalism" on which the whole enterprise was based. Instead, "the construction of history" is to be grasped "as such. In the structure of commentary."[23]

History emerges from the structure of commentary as citation. What makes it commentary is the fact of a citational grafting or montage that begins with the historical object being torn out of context:

decontextualization and recontextualization is its structure, beginning in the form of an absence of structure, or what Heidegger might have called an *Abgrund*. But again, that abyssal *discontinualization* is a coming-to-history as the structure of life, for it is the means by which something flashed up comes to have the "history of its own" on the basis of which it "is credited with life" (*Writings 1*, 255). To write history by means of a montage of citation and commentary, commentary as citationality, is to produce life as inanimation.

For the life of that written, cited history includes a disjunctive effect of a different kind, namely, the automation that we discussed in previous chapters. Citation, as a function of repeatability or iterability, makes explicit the possibility of mechanical repetition and introduces a lifeless machinic impulse onto the historical scene. In materializing history by citing it, Benjamin's historian removes the event from the continuum to the extent of making it a cog in a machine, contriving it (from Fr. *controuver*, which literally means "discovering together," putting together or juxtaposing to the extent of reinventing) in a context that obeys a whole other set of operations, very different instructions for use. A cited history, a piece of it "torn from its context" and extracted from the continuum, no longer retains its supposed natural link to its origin. The relation between that piece when first mentioned and the same piece once it is cited comes to be determined as a relation of "pure" repetition. Irrespective of how the repeated piece blended into its original context, or folds into its new context, what makes the recontextualization possible is the structure of repetition and the irreducible element of automatism that goes with it. In other words, beyond the interventionist mechanism of extraction and resituation effected by the historical materialist, beyond the limitless number of recontexualizations that are rendered possible, indeed underwriting those very interventions, is the automatism of iterability, the fact that, in being cited, something is repeated, and once it is repeated, the structure of repetition becomes irreducibly inscribed within it. It no longer simply repeats itself without also repeating the structure of repetitivity that renders its repetition possible. That structure of repetitivity remains constant throughout the very differences that each repeated instance institutes, effectively reproducing itself automatically, as if with a life of its own.

Thus, in spite of Benjamin's relegation of criticism as the converse of his promotion of translation, we would have to situate any form of commentary in the place of the angel, opening the future dorsal space that was always there. The critic, indeed, any reader finally, also disjoins the work, puts it out of joint with respect to itself, interrupts its temporal continuum—and technologizes it.[24] Not that the work, for being produced as artifact, was not itself a technology; indeed, not that the human that produced it was not itself a technology, never existing in any pure, natural state exempt from or immune to articulations with otherness. But, in a movement very similar to that whereby the angel transforms nature into history, it is thanks to what we call criticism or commentary, analysis or exegesis, interpretation, explication, or simply reading, that a certain secure or intact circuit of naturalness about the work comes to be ruptured. Perhaps the most explicit form of that rupture is what might be perceived as the usurping of authorship by the critic, but one could also understand it via the perversion of a supposed natural, noninterfering environment of reading or via the industrialization and mercantilization that the professing profession and academic institutions of textual studies constitute, however much we might wish to have it otherwise.

A juxtaposition or montage of the task of the translator who rewrites with that of the historian or critic who cites, such as I have been arguing here, therefore leads us to understand the mechanical functions of materialist historiography, not as lifeless clockwork, not as its death, but rather as the sign of its afterlife, survival, or *survivance*. The technology that arrives with the angel, on or as its interface, at once in front of and behind it, the technological force or storm of progress, is what, by interrupting, repeating, and transforming life, makes life, revives life, produces life after life. To the extent that we are prepared to follow Benjamin's logic of a text that lives on by being translated or critiqued, we have also to conceive of a life that similarly lives on, in, through, and by means of being torn out of its context, losing its link with its origin, becoming automatic, in a sense, inanimate. If history comes to be by being interrupted, if it comes to life by having its continuity shocked out of itself, then life is discontinuous. That means not just that there is a certain discontinuity within life but that *life is defined and structured by* that very discontinuity; it requires us to accept "technological discontinuity" as the very life in life. What gives life, what makes life life in these terms, is its technological repetition, its originary automaticity.

That is what Benjamin enjoins us to think if we are to pride ourselves on not inhabiting one of those "times of narrowly prejudiced thought" to which he refers and accept his "inkling that life [i]s not limited to organic corporeality." That said, we would have to admit that as yet we have little agreement about what that inkling might amount to, even as the inorganic makes such rapid and radical inroads into our corporealities. For our purposes here, however, it might include continued examination of the task of the translator or critic in the context of the originary afterlives that Benjamin describes. It might mean arguing that the critic revives or reanimates a text less by breathing life into it in the manner of the original creator of it than by disjoining it, making something of a ruin of it, which also means *artifactualizing*, and technologizing, but also translating it, less by some ingenuity or some fiat, however, than by somehow recognizing, mobilizing, or activating what is automatically articulating and engaging within it.

Let's see. An image. Picture Benjamin's Klee's *Angelus Novus,* wide-eyed, mouth agape, staring at the past as it is driven irresistibly into the future. Somewhere there between the angel and us, out in front in the past, out in front of this new angel, somewhere else in the gallery of art history, there is another, older *Angelus.* It was painted by Jean-François Millet in 1857–59 (Plate 2).

Millet's *Angelus* doesn't show any angel. The Angelus of its title is a prayer that the peasant couple have paused, interrupting their toil, heads bowed, to recite. This angel is not even a living, mythological, or fantasmatic being. It is rather something *visually* unrepresentable and in fact concerns something that is even *conceptually* unrepresentable. For the Angelus is a short devotional prayer, consisting essentially in the triple repetition of the Hail Mary, which is to be recited each morning, noon, and twilight at the sound of the bell and performed in remembrance of the Annunciation: *Angelus Domini nuntiavit Mariæ* (the Angel of the Lord declared unto Mary). It celebrates Mary's being chosen to bear Jesus. The "Angelus" that comes to be as Millet's painting, or at least as the title of Millet's painting, comes to be as a long series of translations: it is another title, the name of a prayer that cites a scriptural reference to a winged spiritual emanation (Gabriel) that appears to Mary to announce that she is (to be) impregnated by the Holy Spirit—it is the title of the announcement by God's messenger of an immaculate conception by a form of God. As I suggested earlier, and as the classic iconography shows, the conception itself is often understood to coincide with the angel's announcement of it, such that the penetration of Mary's ear by Gabriel's words in a sense performs the impregnation, becomes metonymic to her fertilization by the Holy Spirit. Obviously, very little of that is visually representable, let alone conceptually representable, which is why Western art abounds in attempts at its representation: angels are usually seen as winged humans, whereas the Holy Spirit is either a dove or rays of light, showing an increasing tendency toward pure manifestation that is at the same time a tendency toward unrepresentability.

So that is what we see and don't see in Millet's painting: no visible angel, the center of the visual field instead replaced by a strange emptiness, the silent immobility of the space between the two human figures. In that space, there occurs what we see and can't see: the invisible and mute Angelus, a prayer, wafting up to the heavens.

Now, much of that, it might be argued, is entirely contingent to the image itself. Its title, *Angelus,* less names an angel than it functions as shorthand for something like "Peasant Couple Pausing to Recite the Angelus." Indeed, Millet's original title was *Prayer for the Potato Crop,* relating to a childhood memory of his grandmother's devotion and charity.[25] But without going into debate over how the discursive framework of a painting, including its title, does or does not infringe on or enter into the interior of the work, it is important to note that the reference to the Annunciation given by the title is also an indirect reference to the incarnation. The crepuscular ritual of the prayer, with special ringing of the bells, relates to the belief that it was at that time of the day that the Virgin Mary was greeted by the angel Gabriel. And in announcing to Mary the conceptual conundrum of her immaculate conception, he was also announcing the similar conundrum of God's becoming human.

This twilight of the angel is therefore a complex Benjaminian moment of translation: God is coming to life as man, transitioning from spontaneity to conception in more ways than one; a new Adam is being conceived, who will be required to rename in the sense of instituting a new dispensation, and renegotiate the relation of death to life. Different historical montages are also coming into play: a triune God, a messianic moment for some, new images, precisely, of (in)visibility. Thus I was justified in suggesting that Millet's painting opens up a strange representational space, both from without, by virtue of its title, and from within. That problematic of representation may well have been something Klee was reflecting on when he painted his version, and about which Benjamin is explicit in his theses on the concept of history, for in that space is inscribed the question of materiality itself. The angel that appears absent from Millet's painting quite clearly returns to inhabit its center, via the representational quandaries that the Angelus raises: questions such as what form of materiality is an angel? How is it to be represented? What is an immaculate conception? What is an incarnation? And conversely, what form of materiality is a prayer? How is it to be represented? What is the corporeality of words in general, whether silently recited, spoken, or written?

There appear to be no words in the central visual field of the painting. If we understand the human figures to be speaking, it is in the muffled semi-silent tones of prayer. No thing is allowed to dominate that central visual field: it remains the empty or silent space of prayer. The dominant perspective that is imposed by the form of the two figures and by the space between them—and, by extension, by the prayer, of course—is vertical, leading up, then out, across the horizon to the increasingly brilliant sky of the upper left background. No thing to speak of is to be seen in that direction, and indeed the viewer must resist that perspectival impulse in order to follow the horizon to the right and light upon the church steeple, which was moreover also added after the event, at the same time as Millet changed the title. In that upper right-hand corner, another landscape is constituted by the steeple and other buildings and by the migrating birds above. It offers an abyssal repetition of, or counterpoint to, the dominant narrative of the prayer's ascendance, creating another relay of intercession between the community and the heavenly bodies represented by the birds.

To see *things* in this painting in any strong sense, one has to double back from the steeple, leave the ascending prayer for the birds, come back down to earth and begin to deal with a wheelbarrow, a basket, and a pitchfork. Those solid objects serve as a strong materialist counterpoint to the immateriality of the prayer in its relation to the ethereal sky. They also found the agrarian socialism of the painting, both demystifying and mythologizing the impoverishment of the peasant class. As part of the trio of paintings that included *The Sower* (1850) and *The Gleaners* (1857), *The Angelus* helped establish Millet within the tradition that Zola would consecrate as literary naturalism in the latter part of the nineteenth century. Millet's realism, however, was entirely dedicated to an idealistic chronicling of agrarian poverty. It would take a Salvador Dali to see something in *The Angelus* other than its apparent simplicity and piety. As we know, the surrealist painter had a veritable field day with the middle-ground objects and with the scene in general. In a sense, he wasn't wrong to find sexuality all over, beginning with the handles of the wheelbarrow like open thighs protruding into what might be called the underside of angelic space, insisting that the scene was one of sexuality and morbidity.[26] For what is unspoken here, after all, by the angel and by the couple honoring the event is Mary's uncanny experience of sexuality by means of the Annunciation and Immaculate Conception, an experience that, at a very minimum, had to involve repression of any *carnal*ity within the experience of in*carna*tion. We don't have to be Dali to begin to explore what might be evoked by this intrusion of a wheelbarrow into a sacred space.

What interests me, however, in the intrusion of the wheelbarrow, before any possible erotic force, is its technological *artifactuality*. Whatever there is of the angel, or the angelic, in the middle of Millet's *Angelus*—a whispered prayer, a voice from the ether, the unrepresentable or difficult to represent in general—falls into a brute *hyletic* materiality, wood and iron, a simple thing, simply a thing. The prayer is interrupted in its elevation and brought back down to earth to become wheelbarrow, and Millet's Angelus becomes a ruin at the feet of Klee's/Benjamin's angel, emerging out of the near middle ground of Millet's painting into the foreground, out from behind the woman into the past nearer to us. No longer simply angel, no longer incarnation, if it is to speak now, it will have to do so with the mute force of an object, a wheelbarrow angel. Whatever idea it is or represents will relate to its thingness. For the wheelbarrow angel of Millet's iconic or canonical painting is necessarily part of the history, the art history, at least, that lies before the angel in Benjamin's conception of the historical materialist's task. To the extent that, in Millet, it represents

Gabriel's fall into materiality—annunciation on the way to incarnation and beyond—and to the extent that my reading of that so-called fall is of a necessary and originary technologization, the wheelbarrow angel can be seen to coincide with Klee's image as read by Benjamin. In other words, the image of the *Angelus Novus* that Klee gives Benjamin to read as a visual incarnation of how history becomes dialectical materialism, by means of image and montage, citation and recontextualization, is necessarily open to further translation; indeed, its afterlife depends on that. Such a translative montage might take the angel further into hyletic materiality to produce something of a new life for it. For whereas Dali, and in a sense Lacan after him, seeks by means of his paranoiac criticism something hidden, more or less immaterially, beneath the painterly surface, such as the coffin that he insisted on and whose vague shape he claimed to confirm by X-ray analysis, or else seeks an ambiguity of the trompe l'oeil where one image morphs into another, Benjamin's inanimation as translative montage suggests a simple superposition of one image over another. It would be a matter of blasting the wheelbarrow out of the continuous narrative of Millet's painting, and out of its idealistic piety, to project it as something simple, utilitarian, a rudimentary machine, a wheelbarrow, nothing high-tech about it but a machine that is itself designed for transporting or translating simple humus matter, the earth or autochthonous matter, that is the space of every immaculate conception. That montage would produce, by means of its simple human conception—a wheel, a container, two handles—a technological translation of the ideology of nature and its supernatural transcendence, such as dominates the painting, producing instead a portmanteau image, a catchall receptacle on the basis of which one could also attempt some triage, and haulage, of the pile of historical debris Benjamin describes as growing toward the sky. A wheelbarrow is indeed something his angel needs (Plate 3).

It should be red, for iconic and canonical reasons. For the thing that I have been tracing throughout this exposition as the materiality of the angel is not just a certain wheelbarrow but, finally, a red wheelbarrow— the one so much, which may or may not mean so many things, depends upon in William Carlos Williams's poem. If it were to emerge from the montage of historical images to rejoin the modernist moment, it might look something like Plate 4.

In fact, Williams's Poem XXII from *Spring and All* lies in the very near future, right behind Benjamin's angel. Its publication date is 1923, the same as "The Task of the Translator," and three scant years after Klee's painting. It gives us, it is generally agreed, the thing in all its haiku simplicity: "so much depends / upon // a red wheel / barrow // glazed with rain / water // beside the white / chickens." That thing wants most obviously to be a wheelbarrow, but it is also a poem, and within the poem, words and nonwords, the materiality of both black ink and white space that was extensively discussed in the previous chapter. Following Williams's own maxim "no ideas but in things" (from "Patterson," 1927), commentary on the poem, from J. Hillis Miller to Charles Altieri,[27] has sought to adjudicate the precise level of literality or abstraction represented by the wheelbarrow, to deal with the representational quandary that it poses. Henry Sayre gets perhaps closest to what I am evoking here when he states that "at the center of Williams's poetic is his conception of language as physical material. . . . Williams emphasized the materiality of the poem."[28] As Williams writes in *Spring and All,* "poetry: new form dealt with as a reality in itself."[29] Sayre is positioning himself in opposition to Miller's insistence on a leap into things for their own sake, insisting in a sense that the word *wheelbarrow* (in fact two words in Williams) is a thing arranged in space before the object wheelbarrow is: "so much depends on the form into which Williams molds his material, not the material itself," Sayre writes, before comparing the wheelbarrow with Duchamp's readymade, the urinal called *Fountain.*[30]

As it happens, Sayre's transition is somewhat problematic. If, as he states, what transforms the trivial sentence "so much depends upon a red wheel barrow glazed with rain water beside the white chickens" is its arrangement into four similarly shaped stanzas, then the comparison with finding a urinal and placing it in an exhibition would have to refer to the words, not the everyday object—a wheelbarrow—represented by those words. For Sayre, Williams's material—"the wheelbarrow's accidental but very material presence in this new context"[31]—is ambiguously the object and the word that represents that object. And in any case, both Benjamin and Saussure agreed that a word represents not an object but a conceptual abstraction of that object: there is no material presence of a wheelbarrow in Williams's poem.

In general terms, however, Sayre is right to point to the visual materiality of Williams's poem, for it is again the play of spatialization, or interruption of the continuous flow of prose, that is seen to constitute poetry. So much poetry depends on that, so much so that Williams will also claim that poetry, at least in its written form, is *produced by* the blank space behind it: "There is nothing to do but to differentiate prose from verse by the only effective means at hand, the external, surface appearance."[32] Not that prose doesn't also have blank space behind it—such that every piece of writing is finally becoming-poetry—but as I have insisted, what constitutes poetry is a different play, and a different explicitation of such space. It represents a particular form of materiality behind the materiality of the words and also something of a representational quandary. Indeed, if we go back to allowing that the blank white space of the page represents the silence of an oral enunciation, a pausing of varying degrees of pregnancy, more or less grave or gravid, then that quiet emptiness is also the space, like a sudden gap in the conversation, through which, in various languages and cultures, one says that an angel passes—an angel that, as I have been insisting, is always already a wheelbarrow. Or an angel that is a wheelbarrow at least from the moment that, in mimicry of the Adamic moment, it is named as such and translated as life. The "wheelbarrow" of Williams's poem could not come into whatever materiality it is that it comes into without the becoming-wheelbarrow of such a white angel, without the disjunctive force of that white space; even less could it come into redness, as imagined.

If we follow Benjamin, every materialization, whether object, word, or color, depends on an intervention that he calls history and that I am calling alternately a translative inanimation of life and a technologization. "Before" it, there may have been nature, but that would be a continuum to which we have no access, which certainly has no history, and on the basis of which it would be impossible to effect anything "human" at all, certainly not anything like a politics. History begins once spontaneity translates into conception, once there is that sort of intervention into nature to transform it and produce it as materiality. If, as I have been attempting here, we produce that sort of historical materialist reading of Benjamin's writings, juxtaposing or overlaying "Language as Such," "The Task of the Translator," and "On the Concept of History," we can understand him to be saying that such historical technologization or technological historicization constitutes life not only as autokinetic, capable of displacing itself without intervention, but also as inorganic. As we have seen, he gives a name to that technology: (it is the) translation (of the language of things into the language of man).

We would have therefore to imagine a "spontaneous" technology of translation that, translating spontaneity into conception, confusing the autokinetic with the automatic, is imagined to inhere structurally in nature, to be always already at work in the continuum.

The inanimating technology of translation is perhaps another form that could be traced behind the angel, at once its past condition of possibility and its future becoming. Which might once again make it a red wheelbarrow. First, because there is no language that is not a technology, which makes its relation to the organic and the human as close as that rudimentary machine by means of which a human carries soil—among other things—from one place to another. A carrying across is, as I have insisted, the literal sense of translation, and by means of it the orders and priorities are displaced and transformed, for the soil also supports or carries the human and the wheelbarrow that move across it, and the wheelbarrow also "transports" and transforms the human from one place to another, from home to field, from repose to toil, from use-value to exchange-value, and so on. The second reason to understand it as a red wheelbarrow is because a fundamental technological translation takes place by means of language, most explicitly by means of the organic intervention within language that we call poetry, an intervention that mechanizes language by means of its inbuilt functions and capacities. For it is in the first instance a poem, poetry that speaks and says that so much depends / upon // a red wheel / barrow //; poetry itself depends on the very possibility of a straightforward, as it were, organic, utterance, but one that is at the same time a halting and disjunctive, as if mechanical, articulation.

We can hear, read, or know something of that because we face the writing of the wheelbarrow as we face the angel. But we can get out in front of it and become the future harbingers of the full sense of it only from a dorsal perspective, taking into account the workings of it—words, blank space, translations of one into the other—there where the wheelbarrow angel passes headlong into its history to come.

6

RAW WAR

Schmitt, Jünger, and Joyce

No raw history, Benjamin insists. So then, raw war? War, raw as it was always supposed to be, a naked physical contest, a Pancratic *corps à corps* that barely crosses the line separating it from sport? War, raw as it should be, an even-handed combat with mutual respect for rules of engagement, and a symmetrical separation of good on one side and evil on the other? Raw war: the presumptive phantasm of a warrior ideal that persists well into the most sophisticated contemporary military operation?

Or instead: war that is, before any obvious political impasse, a contentious fault line running through language in whichever direction one chooses, and of which the truculent call to arms is merely the most transparent symptom; war whose front gets turned around, extended, diluted and divided by what lies behind it; and war that is in essence rapacious victimization and tortured flesh.

War, raw to the extent that it is fought at the extreme edge of life, where life reveals its inextricable relation to the technological.

Carl Schmitt seems to have taught us that it is war that creates politics, rather than vice versa: there is no enemy without the "ever present possibility of combat"; war must remain "a real possibility" for the concept of the enemy to be valid.[1] War is thus the condition of possibility for the concept of the enemy (as well as of the friend), which, in Schmitt, is the condition of possibility for the concept of the political. In turn, however, war relies on a concept of the front. For Schmitt, one has to presume a homogeneous spatial extension circumscribed by a frontier, a line along which an opposing force, operating at the outer limit of another homogeneous spatial extension, will be drawn up. Thus it is the front, finally, that defines the war that defines the enemy that defines the political.

Now, such a logic might be little more than a self-evidence that deflates for lack of substance: (international) politics begins with a border between nations; or it might be a horrifying presumption concerning perpetual war: neighboring states can enter into political relations only as long as the frontier between them is a line that divides friend from enemy, and only as long as that frontier is subject to contestation by means of war. To repeat: a given spatial or geographical entity can constitute itself as political only as long as it is at war, potentially or actually, with a proximate enemy, from which it is separated by a front(ier). And that same frontier must as a result be, potentially or actually, in dispute. The paradox of Schmitt's logic—one among others—would be that the supposedly fixed anchoring point of the concept of the political, namely, the front as precise site of combat, is subject to constant shifting under the threat of, or as a result of, that combat; it is constantly redefining the spatial homogeneity it is presumed, by delineating it, to circumscribe.

We can perhaps identify there the basis for another apparent contradiction, between Schmitt's insistence on the enemy as public rather than private yet at the same time "existentially something different and alien" (*Political*, 27). An enemy should be someone with whom one has no personal relation, living far away, over there, but who is nevertheless known to be other, alien, different, and who would presumably have been identified as such on the basis of a confrontational face-to-face comparison—someone seen up close, eye to eye, but who remains an unknown stranger. That is to say that Schmitt presumes politics to derive, on one hand, from an opposition between two homogeneous national blocs that function as public enemies, such that someone far from the front and having no experience of the enemy could still be mobilized against such a foe, and, on the other hand, from an up-close encounter along a contested frontier, on the basis of which one defines the enemy as alien, and because of which one is constantly involved in the business of drawing a line of distinction between friendly neighbor and alien foe.

We can more obviously understand the paradox of the shifting frontier as the reason for Schmitt's awkwardness regarding two other cases: first, civil, partisan, or guerilla wars, where the front dissolves into fault lines running through the supposed homogeneous territory of a single political entity, and second, *teletechnological* hostilities or remote-controlled wars in general, where fighting is no longer necessarily concentrated in face-to-face or, as it were, hand-to-hand combat. For teletechnological combat would increasingly constitute the contemporary polemological norm, inaugurated in one sense by the motorization that Schmitt observed in World War I but emblematized more clearly by the ballistic arc in general, whether of the Norman arrows of 1066 (and whatever precedents already existed, for example, David's slingshot) or the intercontinental missiles, starwars shields, drone bombers, and robot soldiers of the present and future.[2]

Teletechnology extends the front and irremediably problematizes Schmitt's political. Because of it, Derrida suggests, "the political no longer has a place . . . it no longer has a stable and essential *topos*. It is without territory, uprooted by technology, by the unheard-of acceleration and extension of telecommunicational distances, by irresistible processes of delocalization."[3] But we should understand such an extension of the front to have already begun in the relation between the human and its weaponry, indeed, in the outstretched arm of the first combatant, for there is no technology that is not a teletechnology in the sense of a prosthetic extension or redesign of the animate. Between the initial technologization of the human by means of his being armed, through motorized transportation of soldiers and weaponry along or behind the front, to the most delocalized forms of contemporary warfare such as drone aircraft operated by pilots continents away, there are simply (though not simple) extensions of that originary prosthetic structure, which is the structure of extension or articulation itself. Within that structure of teletechnologization, questions are raised that go to the center of the concept of the political. For example, the spatial extension of the front is implied in the question of which of three aims of war is to be prioritized: the elimination of the enemy army, which presumably takes place on a successive series of fronts; the destruction of its arms, which presumably involves multiple sites removed from the front, including the whole infrastructure of a given armament industry; or the occupation of its territory and the pacification of its populace, which means universalizing the front within a given state and often includes incursions into other states. Schmitt tries to be categorical in his choice among those aims: "war is the existential negation of the enemy" (*Political*, 33). That is why one cannot wage it, for example, purely on economic grounds. But that doesn't prevent us from imagining a war fought on the basis of firepower alone, an exchange between missiles and interceptors taking place in a "neutral" space such as the exosphere, with no loss of human life, a war that would nevertheless bring the enemy to its knees by allowing whichever side didn't run out of weapons to emerge as victor. For the vanquished side would thereby be utterly exposed and hence subject to existential negation even though no killing took place. One must henceforth also imagine, more bloodlessly still, perhaps, a purely cyber war, whereby one side renders the enemy's weapons operating systems inactive before those weapons can be unleashed.

Such versions of combat, however, exclude the very principle that appears to constitute and sustain our understanding of war, namely, the inflicting of physical pain up to and including death, and hence Schmitt's insistence on an extreme enmity that necessitates "physical killing" (*Political,* 33). The "enemy concept" is based on an idea of combat from which "all peripherals must be left aside . . . including military details and the development of weapons technology" (32). Conversely, then, it should be possible to imagine a war fought on the basis of pure physical engagement, an unarmed combat, something like a wrestling match to the death undertaken by two champions, by two armies, or by two populations.[4] But it would be naive—and, as I have just argued, conceptually inconsistent—to suppose that the opposition between a war fought purely between weapons systems, and one fought purely between bodies, was an opposition between a technological and a nontechnological war. Again, the "arming" of the body for offensive purposes—or for any other purpose—begins with the training and development of the body itself, its instrumentalization or indeed technologization, a technologization that we would, of course, have to understand to be in effect, even "before" that, within the body itself. The technology, or at least *technics,* of combat does not wait for the invention of this or that type of weaponry to set itself in play.

Indeed, war is necessarily posited on the development of technology, to the extent, perhaps, of being its condition of possibility. Can we determine whether what we traditionally think of as the first technological inventions, rudimentary tools such as sharpened stones, were conceived and produced for aiding in cultivation rather than for use as offensive weapons? And the use of such weapons obviously involves putting a distance between the opposing bodies that wield them—a teletechnologization, therefore—enabling the body of the attacker to act at a distance as much as to protect that body from the opponent's aggression. In contrast, it is difficult not to understand Schmitt's front, and thus his combat, and thus his politics, as a nostalgia for an aristocratic homoerotic form of physical competition, something decidedly Spartan in its conception and decidedly symmetrical in its execution, beginning with a handshake even as it ends with physical killing. Schmitt imagines a war of existential negation and physical killing, posited on a *confront*ation from which "peripherals" such as weapons technology have been left aside. But his war cannot avoid being an "armed combat" fought with weapons whose "essence . . . is that [they are] a means of physically killing human beings" (*Political,* 32–33). Thus the technological necessity is introduced as soon as it has been discounted, even if its distancing or "peripheralizing" effect remains unacknowledged: Schmitt presumes there to be weapons whose essence is killing humans but whose technological development remains peripheral to their use. Acknowledging the technological necessity would, after all, mean accepting that the front is a confrontation between weapons rather than bodies, that however crude such weapons be—and whether the arm be some form of *offensive* sword or *defensive* shield—they are what one leads with into the fray; it is those weapons that perform the physical killing. The weapon of war is in that sense itself a front whose back is the soldier wielding it; humans are increasingly behind the military technology they unleash, behind both in the sense of more and more remotely producing and controlling it and in the sense of being removed, by their weapons, from the front, the front of the front, where it has its effects.

As long as existential negation and physical killing remain the objects of war, a cruel calculus will come into operation whereby technology provides a surplus of life to the winning side. And increasingly, as teletechnological advances allow the living to remove themselves from the theater of war, that calculus will function according to a different type of perversity. It will not be a matter simply of saying that the lives on our (winning) side are protected by our superior technology, that our technological offence defends and saves our soldiers (and wives and children), in the same way that our other technological advantages—for example, medical technologies—prolong and save our lives whether military or civilian, whether in war or in peace. For the surplus of life that technology gives our side is predicated on a loss of life on the other side. More of the enemy must die, and the superior technology that protects us does so not only simply by saving us but also by killing them.

But neither does that mean a simple economy of sacrifice is in force whereby our enemies are offered up for our survival. That is so in the first place because no economy of sacrifice is simple. The loss of life on the sacrificed or conquered side accrues to the victorious as *the surplus of their survival.* Not only do the victors live, they survive: it is, as it were, a different form of life that accrues to them. Their life is enhanced by their being permitted to continue living proportionately as life on the other side is lost, but that does not take place according to a rationalizable ratio. Second, a simple economy of sacrifice is not at work because, in this life-for-life exchange, there operates the hidden "labor" of technology, to which the living on the victorious side and, conversely or perversely, the dead on the losing side are indebted. The survival that is gained on the one side, as much as the life lost on the other side, is a technologically inflected life. Technology has become in this sense not only the messenger and purveyor of death that we often take it to be but also the carrier of life. By acting at a distance, by acting as the front in this exchange of life and death, by performing combat and existential negation on behalf of and instead of the living, technology has come not only to represent but also to perform a function of living, and not only to *protect* but also to *produce* life.

There is a different front that Schmitt brings into play in his politics of enmity, which threatens to add a whole other layer of complication to his model. It runs between the political concept and other domains that, he considers, can and should be excluded from the discussion, namely, "moral, aesthetic, economic, or other distinctions" that would "weaken" the friend–enemy antithesis (*Political,* 27, 28). Schmitt aims to protect the coherence of his concept from two main versions of such distinctions: first, traditional moral or theological discourses—from Machiavelli to Hobbes, Bossuet, de Maistre, Fichte, and (sometimes) Hegel—whose political theories "presuppose man to be evil" (61); second, forms of liberalism, indeed "every typical liberal manifestation" that gives "every political concept a double face," allowing it to be clouded by "ethical or moral pathos," on one hand, and "materialist economic reality," on the other (71). It is precisely in contrast to the abstractions of morality, aesthetics, and economics represented by such discourses that Schmitt defines the political as the *polemical*; he might therefore be understood to be rescuing, not just the political, but also the polemical from rhetorical abstractness, returning polemic to its literal Greek sense, that which relates to war.

There is thus a literalist and materialist presumption to his notions of the front, combat, and physical killing and, by extension, to his concept of the political, but one whose circumscription, like any such presumption, is difficult to maintain. In the first place, one cannot imagine a war without a discursive apparatus behind it, from the simplest performative declaration of it to the whole dense mass of political rhetoric that forewarns of it, rallies around it, and explicates or justifies it. Whereas Schmitt might base the political on physical killing precisely to determine it as a domain of action rather than words, the logic of his argument is not to reduce politics to combat, for that would negate it, nor does he want war itself to function perpetually as the guarantor of the political: "It is by no means as though the political signifies nothing but devastating war and every political deed a military action, by no means as though every nation would be uninterruptedly faced with the friend-enemy alternative vis-à-vis every other nation . . . the political does not reside in the battle itself . . . but in the mode of behavior which is determined by this possibility" (*Political*, 33, 37). What he requires therefore is rather the "ever present possibility" (32), the "real possibility" (33) of combat. Something between a distant and abstract possibility and perpetual physical killing will provide the basis upon which people will establish who their friends and who their enemies are. But Schmitt's "ever present possibility" will need to function more like an ever present *threat* if it is to motivate citizens to group themselves according to amity on one side and enmity on the other. Similarly, those citizens would have to react with fear to the ever present possibility of combat in uniting themselves as a political entity—or, if not with fear, at least with some combination of the ratiocinative and emotive, some adrenal release, such as one would expect to come into play in a situation of even potential existential negation. The fact is that, however much he seeks to install the combat–politics relation as a simplifying principle, in insisting on the close relation between the extreme experience of physical killing (or dying) and the political, Schmitt has activated a complicated semiotic and rhetorical network, a network relating to life and death, on one hand, and to the politics of war, on the other. It should happen, he thinks, that we look across the way, conceive of our potential existential negation approaching us from over there, that we consequently look for and at our friends around us, and so come to do politics, without any moral, aesthetic, economic, or other distinctions being enacted. But given what gets mobilized whenever there is any possibility of existential negation, one doubts it can happen that way.

There is, after all, a whole mythological rhetoric, and rhetorical mythology of combat—indeed, a whole mythology, mythology in general, and a whole rhetoric, rhetoric in general, starting with the *Iliad,* that takes combat as its necessary condition and dominant narrative instance. As a consequence, there would be no literature without combat, no narrative that didn't proffer the situation of war as either the heads or tails of its tales of love and romance. In a literature that was close to Schmitt in a number of ways, Ernst Jünger celebrated, as we know, the lure of a "yearning for great experience, such as we had never known. . . . The war was our dream of greatness, power and glory. . . . There is no lovelier death in the world."[5] Once combat comes to be the hallowed experience that Jünger understood it to be, it cannot easily be again reduced to the simple literality of real physical killing. Instead, it rejoins religious or amorous experience in terms of its level of psychic and, along with that, rhetorical investment.

We should not expect, therefore, that the soldiers belonging to a given political entity attack and kill their enemies along a line of confrontation on the basis of an unmediated response to the bellicose exhortations of their side. Indeed, even if we were to presume that politicized citizens, and their martial representatives, could be induced to act like such obedient automatons, attacking and killing fellow humans simply because the latter were declared to be enemies, without any other motive or motivation (moral, aesthetic, or economic: "they are more evil, more ugly, more wealthy" or "they are doing us harm, in bad taste, and making us poorer"), we would still have, somehow, to account for their recourse to the violence of existential negation on a scale that far exceeds everyday political experience; or indeed, reading Jünger, we would have to deal with an account of that violence that described it as far exceeding the political, as attaining the transcendent level of hallowed experience. We would have to presume, short of the fantasy of a totally unthinking and unfeeling human military machine for which, moreover, it is doubtful that a concept of the political could have any sense, that the existential negation of an enemy such as Schmitt conceives of is impossible, if not without invoking moral, aesthetic, or economic considerations, then at least without activating some psychological effect, something like courage or fear, on one hand, and a will to violence or cruelty, on the other.

On one side and the other, one is faced with a serious complicating of the political concept in its relation to combat. In that respect, Jünger's *Storm of Steel* might be read, in very selective terms, as a literary application of (or, rather, precedent for) *The Concept of the Political*: "It has always been my ideal in war to eliminate all feelings of hatred and to treat my enemy as an enemy only in battle" (*Storm*, 52). Similarly, Jünger's war is conceived of in almost Pancratic terms, as the arena where an aristocratic few are "encircled by a nimbus of romance, even in the most material of wars, owing to their insatiable daredevilry" (211). But that dispassionate and sporting approach to combat is quickly understood to lay the groundwork less for politics than for an ethics of military *virtù*, dominated of course by courage. Thus, eliminating all feelings of hatred and treating the enemy as an enemy only in battle goes hand in hand with "honour[ing] him as a man according to his courage" (52). A discourse of courage redeems physical killing from murder, violence, and cruelty and elevates it to become war; with the addition of courage, bloodlust becomes calm and invulnerability. However, although it serves as the basis for the rhetoric of combat, courage itself has little semantic clarity. It is both passive, a form of patience designed to exorcise fear, and hyperactive, a recklessness mobilized in the exploit—both endurance and impulsion: "I have always pitied the coward, in whom battle arouses a series of hellish tortures, while the spirit of the brave man merely rises the higher to meet a chain of exciting experiences" (158).

Once courage is associated with "exciting experiences," or even invulnerability, it can no longer preserve its moral integrity and very quickly becomes psychologized. Offensive courage, taken to the extreme, becomes more or less indistinguishable from the destructive drive. Jünger therefore appears far from Schmitt in the heat of combat—which we might still presume to be the essence of the political—for he describes it as a beckoning "carnival of carnage" (308), a political scene determined in its essence by "the overpowering desire to kill": "Every one was mad and beyond reckoning; we had gone over the edge of the world into superhuman perspectives. . . . I was boiling with a fury now utterly inconceivable to me" (254–55). At that extreme point, the soldier is acting under the "spell of primeval instincts" and relies on the pure literality of blood to impose a transcendent clarity over rhetorical imprecision: "It is not till blood has flowed that the mist gives way in his soul. . . . It is only then that he becomes once more a soldier of to-day" (263). Indeed, this confusion between a (perhaps repressed) desire for self-preservation and an "impulse of annihilation" (257), an indistinction that Jünger also refers to as the "godlike and the bestial inextricably mixed" (255), and which finds its expression and confirmation in the knowledge that blood is flowing, that life has no more mystery but is *exposed* as the *explicit* itself, serves as a powerful allegory of what Freud called, in the same historical context, "the activity of our two instinctual impulses, the life instincts and the death instincts" (*SE XVIII*, 48).

Jünger's war journal emphasizes what Schmitt's theory of a political dependent on combat requires: blood. Without it there is no combat worthy of the name. *Blood, finally, is what defines the front that defines the combat that defines the enemy that defines the political*; it is imagined as the zero degree of its perfectly translatable literality, which must also, of course, be understood as the material literality of life itself. We may not know how to define life, but we certainly think we know what it is when we see blood. Blood is the front at its most animate, and trench warfare, "the bloodiest, wildest, and most brutal of all warfare," saves and protects combat on behalf of life: "Of all the nerve-wracking moments of war none is so formidable as the meeting of two storm-troop leaders between the narrow walls of the trench. There is no retreat and no mercy then. Blood sounds in the shrill cry that is wrung like a nightmare from the breast" (*Storm,* 235). In spite of sloshing knee, waist, or chest deep in mud and shit and it, blood is made to represent the "solid" animate core of the front. Without the blood of trench warfare, war degenerates into "an uninteresting butchery" (235), that of various forms of the teletechnological. For just as Schmitt's front is problematized by motorization, Jünger must deal with the paradox of a front that, seen up close, is double, its two sides not in fact conjoined by a gladiatorial embrace but separated by a no-man's-land that is pockmarked by "the mightiest demonstrations of material force" (101), by a "scientific war" that transforms a natural landscape into "the sameness of the machine-made article" (109). Once what he calls "chivalry" cedes the upper hand to machinery in that way (110), once the integrity of the front is called into question, there is nothing to stop war from becoming bureaucratized (285), from degenerating into something that is all distance and technology, a series of actions "ordered from the rear and by the map" (291–92). The front of a chivalrous corps (without or without its horse), locked in mortal embrace with an enemy, gets divided, prised apart by a technology that gives that front a self-extending back.

In such a conception and experience of war, the life lost is life that is lost because, outside of the trench, lives are not lost in the same pure way. A form of life gets lost once blood is shed and lives are lost outside of the trench: what gets lost is precisely the life and vitality of the fatal embrace felt in hand-to-hand combat, the *esprit* of that desperate *corps à corps*. As in Schmitt, it is, in a sense, lost to technology, which, in a variation of the logic outlined previously, must necessarily gain (by) it. Here, however, it is not a matter of suggesting that fewer lives, and less life, are lost thanks to technology. Proportionally, as war succumbs to demonstrations of material force, as it is *materialized* by technoscience, the courage of the fighter fortunate enough to find himself in the heat of the action in the trenches impels him to greater daredevilry. And proportionally, as bureaucratized generals direct a technologized war at some remove from the front, troops are, precisely, increasingly, unwisely committed to the trenches. By keeping their distance, by inscribing a technological distance between command headquarters and the trench, the bureaucrats add to the intensity of the combat and cause more blood to flow. According to this paradox of Jünger's logic, to order from the rear and by the map is to provoke bloodier combat and cause more death; but because blood that flows is the very vitality of combat, that is also to produce more life: more blood, more death, produces more alive (soon-to-be-dead) soldiers. Thanks to technology, then, the lifedeath blood of the trench flows across the cratered landscape, back to the generals' tent. If life is best attained, sustained, even superseded or transcended, in and by close combat to the death, and if technological distanciation both retreats from and commits to that triumph, consecration, or apotheosis of life, then the space constituted by technology—all the way from a no-man's-land of craters dug out by machines to a remote-controlled war-by-the-map—is also a space filled by life, however improbable that may seem. Furthermore, if technology is defined as teletechnology, as a distancing or retreat from the real lifedeath of the trenches, and if—superimposing Schmitt's logic now—the blood in the trenches flows not in vain but as a direct link to the life of the political community that it defines, upholds, and protects, then we can understand it to be technology that installs the very space that is required for frontal combat, on one hand, to define politics, and, on the other hand, for that same frontal combat to keep itself at a safe distance from the political entity that would otherwise be in a permanent state of (civil) war, or else reduced to a composite amalgam of private amities or enmities. The blood of the front must represent the lifeblood of a political entity that it nevertheless keeps as bloodless as possible. If the political is to be kept alive, blood must flow elsewhere. Only (tele)technology can perform the necessary distanciation; hence technology emerges as coextensive with life, from its protection of politicized citizens all the way to the bloody animal or animist ecstasy of trench warfare.

The technological extension of the trench should also be understood as a teletropology. A certain rhetorical arc or parabola, like a gravity's rainbow, bends open the space of the front, despite Schmitt's resistance to any moral, aesthetic, economic, or other complication of the sphere of combat. Whether it be Jünger's blood as the soul's mist-lifter, or the approximation of a map that doesn't show the real lay of the land and, as it were, invisibly sends soldiers in the wrong direction, or the missive-missile that opens adestinational space in general, the contamination that creeps into the supposed necessary relation between politics and combat will take some form of the discursive other of pure action. For, in the simplest sense, Schmitt's combatant is supposed to *speak directly,* indeed, *speak, by his actions, directly for* the political entity. Combat would be that sort of instantaneous fiat by which politics gets unwaveringly expressed. But increasingly, as the relation involves a space of mediation—for example, the distanciation and technologization of the front—it will also constitute a tropologization. We might therefore understand the teletechnological front as a space of *translation,* not unrelated to the originary transformative translation that we encountered in the preceding chapter. For a relation between divine spontaneity and human conception is indeed the context of the war over language that breaks out at the end of II.1 of Joyce's *Finnegans Wake,* in the passage, or rather, the three words, that Derrida has analyzed so incisively: "And he war." That war of words takes place in a space bounded by an arc of *différance,* on several fronts.

First of all, by means of the narration of mock-biblical apocalyptic events in the passage in question:

> Of their fear they broke, they ate wind, they fled; where they ate there they fled; of their fear they fled, they broke away. Go to, let us extol Azrael with our harks, by our brews, on our jambses, in his gaits. To Mezouzalem with the Dephilim, didits dinkun's dud? Yip! Yup! Yarrah! And let Nek Nekulon extol Mak Makal and let him say unto him: Immi ammi Semmi. And shall not Babel be with Lebab? And he war . . . Great is him whom is over Ismael and he shall mekanek of Mak Nakulon. And he deed.
> Uplouderamainagain!
> For the Clearer of the Air from on high has spoken.[6]

We are here far from the clarity of the Edenic scene, where God *spake* and brought things into being for man to name. A mist has descended upon that prepolitical moment without front between God and man, and where language functioned not as a barrier but as a simple discursive relay. Things have instead deteriorated into world war, a universal divine–human combat as a result of man's technological self-extension: we are thus now in the Babelic construction zone and its ground zero destructive aftermath. But the myth of Babel makes clear that once God, as *logos* that was in the beginning, inaugurated discourse, he had already introduced the terms of a technotropological expansion that it would be beyond even his power to control. That is to say, if God is indeed the *logos* he presumes to be, and that the *logos* presumes him to be, the discourse that he inaugurates will remain undifferentiated or *undifferantiated*; it will be the flawless communication of truth, a discourse that is not one, just some sort of limpid transmission of what never needs to be said, simultaneously performing whatever needs to be performed, or creating whatever needs to be created. But, according to the same logic, for Babel to come about, for competition to arise between heavenly and earthly powers, that "discourse" of God, the discourse of his truth and priority, had not to have been understood. The (second) fall, which has as its material and technological fallout the collapse of the tower of Babel, must therefore have already taken place in the first Edenic scene. In those first utterances, as we read Benjamin reading them, there was already more than seamless transmission from spontaneous divine creation to "conceptual" human naming: there was already transformative translation; indeed, life itself depended on such an operation. The judgment of cultural and linguistic difference that God now visits upon the Shems—"the Lord scattered them abroad from thence upon the face of all the earth . . . the Lord did there confound the language of all the earth" (Genesis 11:8–9)—must already have been a fact or situation from which they were precisely attempting to protect themselves. "Let us make us a name, lest we be scattered abroad," they affirm (Genesis 11:4). Let us get back our conceptive creative powers, get back to naming, but let us begin by working for ourselves, begin by giving ourselves a name, which is what we, primary animals, should have done the first time around. In so saying or thinking, the Shems sought to preempt the fact of a translation and transformation of God's spontaneous uttering into their conceptive naming, a translation that had already taken place; they sought to preempt the fact of there being, at a minimum, already two languages. And if it is rather a reputation than a proper name that they want to make—"let us be powerful and stand united lest others exploit our weakness"—if it is the nickname Number One that they want to give themselves, then the fact of linguistic and tropological drift is all the more affirmed: the Shems are necessarily scattering themselves abroad already, by means of their nominalizing expansion. In that sense, Babel is the *après coup* symptom of originary trouble with an undifferentiated *logos,* its revelation as *différance.* It inscribes in an explosive way, and in a pugnacious manner, the opening of translation within discourse, the space of its difference that allows language to mean and allows rhetoric to flourish.

In the second place, or in other words, as Joyce *lets slip* with the sentence "And he war," God's self-imposition, his lording over the builders of the tower, the means by which *he was* the I that I am, is, as the context of *Finnegans Wake* again bears out, quite literally a declaration of war—not just of a particular war but of war in general. God declares war on the Shems and gives them their comeuppance by sowing the confusion—cultural and linguistic difference—that will, in one way or another, be the basis of every subsequent conflict. God inaugurates war both constatively ("you will henceforth differentiate yourselves on the basis of language and ethnicity and deal with the conflicts that arise as a result") and performatively ("whoosh!—did you hear that tower fall?—it was the sound of war(s) to come without ceasing, get used to it").

But, contrary to what we—and no doubt Schmitt—might presume, wars do not in fact arise because there is too much difference, too much of a separation between cultures, religions, and languages, a clear and distinct front of incomprehension and incommunicability, but because there is too little difference, because there is *différance* conceived of in its most straightforward sense, ruination of absolute difference, bilingualism and corruption of linguistic homogeneity, cross-cultural and cross-frontier commerce and intercourse, intermarriage, and so on. That seems evident in the case of so-called interethnic wars and border disputes of the ancient or contemporary variety—from Ephraim-Gilead to ex-Yugoslavia—and it both confirms and subverts Schmitt's classic conception of the front as frontier. And, paradoxically once more, where Schmitt's combat between two alien entities desiring each other's existential negation does appear to function most logically, namely, in wars of conquest and imperialist expansion or punition from Carthage to Iraq, the concept of the front is at its most fragile, subject to dissipation, for the invading side departs far from (or progressively further from) its own borders to transgress a different frontier, producing a front where there was none, and only long enough to perform an ad hoc rewriting of the map.

It is the fact of too little difference that renders impossible, and so radicalizes, any attempt to preserve the purity or unity of this or that ethnic entity or alignment, however real or phantasmatic it be to begin with. God's stroke of belligerent genius consists less in requiring humankind henceforth to deal with the Babelic confusion of different languages than in imposing the Babelic insecurity of imperfect communicability, whether between and among languages or (as the Shems seem not to understand as they construct their direct vertical line of communication to the heavens) within any given language. *Finnegans Wake,* of course, stages such a confusion both in the familial conflict that comes to a head in this section of the novel and in its linguistic performance in general.

Third, therefore, not only does war break out because of linguistic and other such differences, it also breaks out *in* language. That happens as soon as something like "and he was" comes to be written as "and he war." For the sentence remains intelligible in two ways or on two sides: on one side, for example, it is both an "and he was" that has slipped into German and a portmanteau that condenses the verb "to be" and the verb "to war"; on the other side, at the same time, it installs and retains within itself an irreducible element or structure of incomprehensibility. Any war could be understood as a function of that phenomenon, not just, I am suggesting, in its metaphoric sense—not just as the specifically linguistic version of irreconcilable difference(s) but because the ruination of absolute difference, the necessity and impossibility of translation as a problem of the frontier and therefore as potential war, is, in the first instance, or one of its first instances, a fact of language at the beginning of culture. Translation is the technological doubling of language that takes place across a frontier. It is the originary *différance* of language made manifest, materialized and as though mechanized. Once there is translation (and, to repeat, there is translation as soon as there is even one language), there is semiautomatic doubling, whether a machinery of words substituted rapid-fire or speed-of-light one for one or a sophisticated rhetoric of approximation. Such an inscription of the front within language and between languages figures the beginning of the mortal combat that will divide friend from enemy and define the political in Schmitt's sense. Translation means that there are, coextensively, series of human agglomerations divided within and without by linguistic difference, en route to becoming political entities. And although a "precultural" or prelinguistic being might look across a natural barrier and decide to attack another such being on the other side, that simple atavistic aggression would not constitute a political act of the level of conceptualization that we surveyed at the beginning of this discussion; for there to be politics in Schmitt's terms, there would have first to be some cultural and linguistic coalescence. Thus, if Schmitt were to analyze fully the dilution of the front that he laments because of motorization, he would need to look at the problematization of the concept(ion) of the political as a question of translation in and of language. The shifting or divided front might then be understood as less a simple technological expansion of the trench than the originary *translative* or transformative technicity operating inside the very concept of the political, at least as soon as it requires translation across a chain of significations: concepts of the enemy, of combat, of the front, and of blood that flows—and no doubt before the very concept of the political emerges, that is to say, in the concept of the concept.

As we have consistently maintained, language, as both mechanical instrument and conceptualizing function, fundamentally inheres in the technologization of the human. It inheres in the hominid evolution, in groupings into some form of community, such as will eventually emerge as a *polis* and lead to building a tower. And the tower of Babel is not just an af*front* to God but also a frontal provocation in the sense, already mentioned, of an imperial challenge to surrounding communities: "Go to, let us build a city and a tower, whose top may reach unto heaven; and let us make us a name, lest we be scattered abroad" (Genesis 11:4). The *construction* of the tower is, as the Shems explicitly state, simultaneously the *production* of a name, one that functions as the cohering force of unitary homogenization. Prevention against being scattered abroad is assured both by the name and by the tower, two functions of a fortress mentality on the basis of which everything and every one not so cohering will be perceived as seditious and inimical. The tower marks a specific instance of homonization as technologization, which God must punish not only because the front is in the first place drawn in the heavens and in God's face, nor simply because the Shems' affront takes the form of a superior and hence blasphemous level of cultural and technological sophistication, but also because the somewhat slow and dim-witted God that he is *that he is* comes to realize that when he created this race of bipeds, he created them as technological animals, astride the frontier between animal and god, or beast and sovereign, originarily at war with and within their prosthetic selves, constituting originarily a "biosynthetic alloy" (*B&S I*, 80).

This is not, as we know, God's first geopolitical blunder. Supposedly, however, the Shems' effrontery didn't bring to an end the Noahic covenant, instituted after the failure of the Adamic covenant. God went down, confounded their language, and scattered them abroad, but life went on. He did a little precision bombing without provoking a major diplomatic incident. This was not the fire next time implicitly threatened two chapters previously, when God promised no more floods to destroy the earth and produced a rainbow to seal the pact: "I do set my bow in the cloud, and it shall be for a token of the covenant between me and the earth. And it shall come to pass, when I bring a cloud over the earth, that the bow shall be seen in the cloud: And I will remember my covenant, which is between me and you and every living creature of all flesh; and the waters shall no more become a flood to destroy all flesh" (Genesis 9:13–15).

The rainbow is, however, quite literally, a meteor ("*Meteorol.* Any atmospheric phenomenon, as precipitation, lightning, a rainbow, etc."[7]). Between chapters 9 and 11 of Genesis, between the rainbow following the flood and the destruction of the tower of Babel, God made what, for god-watchers, would have to be interpreted as a major diplomatic pronouncement. He installed two types of fire in the sky, both the ballistic arc of the surface-to-surface projectile and the as-if straight vertical trajectory of the air-to-ground missile, and warned of his prerogative to have recourse to either. Narratively speaking, he refined his technology and thereafter distinguished more clearly between living creatures in general and those constituting his political enemy, between nonbelligerents and belligerents. He abandoned total destruction in favor of the precision strike. But when the flood is considered in the context of his subsequent attack on the World Speak Tower, one is forced to understand that every vertical blow has its arc of destruction, that there is always a collateral radius of death. The tower of Babel was no doubt indiscriminately a surveillance center and a citadel, a place of worship and a university. Or even if it were simply a tower for a tower's sake, being built by a well-paid military construction team, one can be reasonably sure that the dead were not all combat personnel. The communiqué says only that "*they* left off to build the city" (Genesis 11:8, emphasis added), but we can be sure that not even God was able to avoid civilian casualties.

The rainbow does therefore, in a sense, deliver on what it promises: better technology, less indiscriminate killing, military precision, reduced collaterality. But more than that, technology functions within its own meteorological arc, in a sense as naturally as a rainbow. Technology is that very arc of natural precipitation, or at least the refraction of such a natural precipitation out of itself. That can mean a number of things. First, tautologically, technology *is*, in peace as much as in war. Second, banally, technology saves life as well as it destroys life, whether in peace or war. Third, mathematically, precision technology may be capable of reducing civilian casualties in war, reducing the collateral radius, but it requires, as a result, a calculus that is equally precise in its cynicism, measuring the value of a particular enemy target and, by extension, the value, in terms of expendability, of those to be found near that target, whether an associate, wife, or child. Finally, conceptually, therefore, high-technology warfare is, by definition, no less than carpet bombing, a wanton breach of the front between combatant and noncombatant.

The technology of war describes its own arc of death and destruction. Within it, there is more or less death and destruction; outside it, life continues, remains, survives (albeit it in another technological arena with its own arc of death, and even, of course, in a putatively nontechnological nature where, thanks to lightning, rain, and all the rest, nature is capable of "technologically" exceeding itself with fatal results). The receding or expanding technological arc puts into effect its own shifting front, between the death within it and the life surviving outside it; in that no-man's-land, there is survival, or *survivance*, thanks in one sense to the precision of technology, but also in the sense of a living that technology defines and produces. Where life goes on is not just where technology has left it to be, or helped it to be prolonged, but where the front between life and technology is recognized as the fiction that it always was. It is not just that technological precision recedes or expands but also that a life that was always already a negotiation with technology expands or recedes in converse. The life that lives on on the threshold of technological destruction is a life that survives at the same time beyond, by, and in technology.

To Joyce's letting slip "and he war" might be juxtaposed the earlier *lapsus linguae*, if that is what it is, recounted by Freud in *Beyond the Pleasure Principle*, when the "good little boy" we understand to be his grandchild

Ernst makes it "quite clear that he had no desire to be disturbed in his
sole possession of his mother":

> Throwing away the object so that it was gone might satisfy an impulse of
> the child's, which was suppressed in his actual life, to revenge himself on
> his mother for going away from him. In that case it would have a defiant
> meaning: "All right, then, go away! I don't need you. I'm sending you
> away myself." A year later, the same boy whom I had observed at his first
> game used to take a toy if he was angry with it, and throw it on the floor,
> exclaiming: "Go to the fwont!" He had heard at that time that his absent
> father was "at the front," and was far from regretting his absence; on the
> contrary he made it quite clear that he had no desire to be disturbed in
> his sole possession of his mother. (*SE XVIII*, 15)

"Go to the fwont!" is Strachey's inventive translation for *Geh' in K(r)ieg!*
Much could be made of this shibboleth in the form of a rhotacism (whereby [r] becomes [w]—"*w*ound the *w*agged *w*ocks the *w*agged *w*ascal *w*an"
was the ditty my father mocked me with when I was a child, sufficient
to tease me out of my impairment by means of a gentle shock-and-awe
therapy) occurring on the frontier between German and (mostly) British
English. Much could be exploited among the different technological effects
it generates as, on the English side, it comes full circle from the Nordic
and Germanic via Scottish and Northern English, across to the American
eastern seaboard and into the wide gamut of variants of it found among
different historical and ethnic communities in the United States. A speech
impediment (I am consciously using the term that carries military bag-
gage [Lat. *impedimenta,* "army supply train"]) must, after all, represent
the fall into technology of (an always already technological) language. By
means of it, language ceases working naturally, breaks down, becomes
mechanically defective, requires orthopedic intervention.

On the side of the English, then, as Strachey interprets it, an explicit
defensive strategy transforms "front" into "fwont." On the German side,
as practiced by little Ernst, we can hear a more passive omission of the
troublesome consonant, a form of neutrality perhaps. Between the two,
between "Kieg" and "fwont," some sort of front remains to be identified.
On one side and the other, that front would extend from the home hearth,

where mother and child jealously guard their supposed uninterrupted intimacy, waiting patiently on some verge of the political, all the way to where the father is waging war in earnest and where Ernst wants to consign all usurpers. The war at the front, the war of the front, necessarily becomes little Ernst's war, if for no other reason than because every war combines an atavistic right to self-defense ("nothing in the present Charter shall impair the inherent right of individual or collective self-defense if an armed attack occurs against a Member of the United Nations") and some version of the oedipal drama. "Go to the fwont!" cries Ernst, or else "Go to waw!" *and he wars,* against all prospective interlopers. He *wars,* but his tongue breaks, his language technologizes, itself and the front, extending and dividing it. Jünger's trench, Schmitt's front, full German–English confrontation, stretches to breaking point. The front loses its outer limits. Germanenglish is Englishgerman, Ernst's k(r)ieg Strachey's fwont, the consonantal pickets uprooted, an *r* on one side a *w* on the other—a palindromic symmetry that is both frontal difference and dorsal extension; rudimentary life, stripped bare, reading back as its technological reflection: war, raw.

III

RESONANCE

7

BLOODLESS COUP

Bataille, Nancy, and Barthes

We find a counterpoint to Ernst Jünger's homosocial communion of warriors in Georges Bataille. His *Blue of Noon* is written within the same general European historical framework (1915–35) as *The Storm of Steel*, a time frame during which, we could say, Westphalian sovereignty convulsed for one of its final times. It traces the itinerary of protagonist couple Troppmann and Dirty from their passionate confinement in London, to Barcelona, where the Catalan uprising comes to its bloody conclusion outside their window, and finally to Trier, Marx's birthplace, where they unleash their necroeroticism in a grimy experience that takes place above a candlelit cemetery and at the conclusion of which Bataille's narrator says that Dirty "made me think of soldiers waging war in muddy trenches."[1] The soil of their frenzied coupling—"we fell onto the soft ground and I thrust into her moist body like a well handled plough drives into the earth.... My hands drove into the earth: I unbuttoned Dorothéa, soiling her clothes and breasts with the fresh earth that was stuck to my fingers.... She kissed my naked groin: the earth was stuck to my hairy legs: she scratched at it to remove it.... We were no less excited by the earth than by the nudity of our flesh"[2]—contrasts with, while complementing it, the blood in Jünger. It is on one hand the dirt of their transgressive sex to match the blood of Jünger's carnival of carnage and, on the other, the elemental soil of their union—their being wedded to and bedded in the earth—to match the unifying blood of war flowing through trench and body politic alike. Thus, despite the reference to muddy trenches (and other references that suggest some fascination with fascism), this lovemaking functions in general as a resistance to a politics defined by combat and by fusional nationalism and enthusiasm, whether on the right or the left; it functions as an example of "excremental" expenditure that flies in the face of, and seeks to radicalize, traditional forms of political action. But because of that reference to trenches, this sex reminds us that whether the experience of the limit takes place via mortal combat or love, it is articulated, by comparison or by contrast, and with comparable paradoxes to those we saw previously, through figures of blood.

Perhaps grimy sex isn't love. And indeed, the point I really wish to emphasize by juxtaposing Bataille and Jünger is that, at one extreme or the other of human experience, the very humanness of both sex and love is defined by blood. Bataille, by preferring dirt, seeks to invert the tradition of love as abstracted desire that is grounded in the figure of a heart swollen with blood. But in my view, he cannot help but reinforce, at least by implication, a concept of love that, like war, relies on blood, blood pumping through a beating heart—or two—a fluid mingling that is, in its dominant instance, a melding of two non- or apolitical entities, a private fusion, therefore, that implies a retreat from political space. Whence the common dismissal of it as a naive, self-indulgent, or inauthentic escapism; whence the presumption that its paradigm is found in romantic experience. Indeed, we cannot imagine better employed hearts than those in one of the most exemplary texts of the Western modern canon, Goethe's *Sorrows of Young Werther.* Early in the text, Werther explains his propensity for cardiovascular exertion: "You ask me if you should send me my books?—My dear fellow, I implore you, for God's sake, do not bother me with them. No longer do I wish to be guided, excited, stimulated; my own heart storms enough in itself. What I need are cradlesongs, and I have found plenty of these in my Homer. How often do I lull my rebellious blood to rest, for you cannot imagine anything so erratic, so restless as my heart."[3] And toward the end, shortly before Werther's blood-soaked suicide, the narrator describes the overheated state of Lotte, Werther's love-object: "Lotte had slept little that night; everything she had feared had happened, in a manner which she had neither anticipated nor imagined. Her blood, that usually ran so innocently and lightly through her veins, was in a feverish tumult; a thousand emotions tormented her great soul."[4] In those descriptions, and in any number of other examples that one could cite from literature, high and low, poetry and prose, the heart overflows to bursting as a measure of the pure life that love invigorates. The body becomes all heart (and blood), which means all inside, all organic. In love, the heart is presumed to swell and beat all the louder within the self-enclosed interiority of a human self to unite all the better with another such self in a fusion that is a type of flawless *trans*-fusion: two hearts beating as one, a blood bond constituted by an interpenetration and mingling of fluids.

Such an organic conception of love invites analysis—analysis, first of all, of that very organic naturality. Love, like combat in its Spartan essence, such as Jünger extols, is understood to be fundamentally natural and human, pre- or atechnological. All it requires is two bodies, two hearts, two souls. However, as we have argued, the armed combat that requires human limbs to extend themselves by means of a prosthetic relation with an artificial, offensive or defensive arm is but a continuation by other means of so-called *un*-armed combat, which involves the articulation of the natural arm in a protoprosthetic gesture of offense or defense. So too the caress of the hand or the heart: first, there is the explicit technoerotics of any scene of lovemaking, even that involving two naked bodies and their instrumentalized anatomical parts; second, as I'll develop here, love involves an exterior articulation of the heart, a peculiar inanimation that is both its swelling, as life is lived to the utmost, and its rupturing by means of exposure to radical, even inanimate otherness.

One might begin to analyze that traditional figuration of love as swollen heart by examining the paradox of its supposed unmediated literality. In love, again as in war, there is a presumed reduction of the tropotechnological arc or parabola of rhetorical effect. In the case of war, the reduction enables the lifeblood of a political entity to flow all the way to the front where it is shed, while at the same time remaining within that entity's veins. Schmitt's combatant, as we have seen, supposedly speaks directly for the political entity, and actual "physical killing" is putatively the unmediated expression of an irreducible enmity between two political bodies that confront each other at the front where war takes place: no more time or place for talking, just doing it in the pure action of bloodshed. Blood and soil are the signifiers of that literality; my blood, and that of my enemy, flows into my soil or his soil, whose ownership he and I dispute. In the case of love, physical union is presumed to express a converse literality, an unmediated communion of caress and penetration, the fluid melding I just mentioned, and a discourse reduced to pleonastic repetitions of "I love you," or something even less articulated, the groan and inaudible grimace. But, like war, love in fact gives rise to a limitless discursive and rhetorical expansion: the lyricism of a Jünger and a Bataille, or of a Homer and a Sappho, and everyone in between, responds to the excesses of a Céline or a Goethe. In the very place where discourse is seen to reduce to its most elemental expression—blood that flows, bodies that merge—we observe an irrepressible contagion of rhetorical production; in the very same place, we find language that overflows itself with itself, as if for its own sake.

A purely literal heart, a heart that was nothing but heart, would be what Jean-Luc Nancy calls "a heart of stone."[5] If we could conceive of a heart that was just a heart, we would know what literality was, we would have found it—dare I say found its heart?—but the heart we would have found would be a stone. It would be solid and immobile, the sort of secure center we would expect to find at the heart of things, but it wouldn't be a heart such as we expect a heart to be, beating to regulate the circulation of blood, engaged in an energetic network that extends throughout the body. Besides, to say that we would have found a heart of stone at the heart of things is also to presume that we were first able to conceive of a stone that was just a stone. That would mean a stone that was not even, for example, a stone that one couldn't get blood out of, for a bloodless stone is already—thanks to a series of figurations beginning with a comparison (it is like a heart or body drained dry) and ending in an opposition (it is a heartless heart)—something of a heart, a heart that is not a heart. Nancy's emphasis, in "The Heart of Things," is in the first instance less on the status of the heart than on the "thingness" without which thinking itself is impossible: "One can think nothing without thinking this inappropriable property of the thing, and without thinking it as the heart of thought itself" ("Heart," 170). However, if thinking cannot advance without first grounding itself in the concept of the thing, neither can it advance by remaining at what I am calling that literal level, at the level of the thingness of the thing. Such static ratiocination would not yet constitute thinking: "Thought finds itself at the heart of things. But this heart is immobile, and thought, although it finds itself there and attunes itself to that immobility, can still think itself only as mobility or mobilization" (171). Having found the solid stone of thingness on which to base itself at its heart, thinking, in order to function as thinking, would have to be able to move that heart of stone, enliven it or have it beat. It would have to loosen or desediment its inorganic mineral compactness, what Nancy calls its "concentration," as if to produce a ripple or transmission of centrifugal concentric effects that would effectively unpack and expose its otherwise motionless, affectless, and meaningless heart.

The relay between stone and heart that Nancy produces in "The Heart of Things" can be read as a particular instantiation of what he has more generally called *exscription*. Exscription describes the operation of sense making in general, and it therefore begins with the heart and the stone *in and of themselves,* that is to say, in the very fact of their not being able to function in and of themselves, as immobile blocks of meaning. For a thing such as a stone or a heart to mean in any meaningful sense, it must be named, and, as Nancy explains, "'exscription' means that the thing's name, by inscribing itself, inscribes its property as name *outside* itself, in an outside that it alone displays but where, displaying it, it displays the characteristic [*propre*] self-exteriority that constitutes its property as name" ("Heart," 175). Nancy first uses the term in a short text on Bataille to explain the community he shares with that writer concerning the "pain and pleasure that result from the impossibility of communicating anything at all without touching the limit where sense [*sens*] as a whole spills over outside itself, like a simple inkblot over a word. . . . That overturning of sense that *makes* sense, or that overturning of sense into the darkness of its source in writing, I call *exscription*."[6] In the context of Nancy's essay, exscription refers, in the first instance, to the relation between Bataille's writing and commentary on it, or to Bataille's relation to his own writing and to a certain impotence that he expressed systematically in his work, an inability to say what he meant, to get to the end of writing, and so on. But, as Nancy explains in a February 2000 interview, he intended it—still inspired by Bataille—to relate more generally to the relation between word and thing, and even as something of a reaction against the presumption of a language, particularly a writing (*écriture*) as an end in itself: "It's a word that came to me as a reaction against a whole craze for writing, text, salvation through literature, etc. There is a sentence in Bataille: 'Only language indicates the sovereign moment where it is no longer in circulation.' That is my daily prayer. It means: there is only language, I grant you, but what language indicates is non-language, things themselves, the moment where it is no longer current tender."[7] The *excrit* differs from the *écrit* not by simplifying the relation between word and thing, by saying, for example, that one precedes or trumps the other, nor even by refusing *univocity,* or the unequivocal, in favor of a *plurivocity,* or pluralization of meaning, but rather by more radically unsettling that relation. It is a matter of insisting that sense stumbles under its own weight and always functions in a relation of externalization with respect to itself, as a type of external edge of itself.

Sense, for Nancy, cannot *reside* anywhere. It has no place to settle in(to); but more than that, it has no self-sufficiency or integrality that would allow it to settle somewhere as itself or in itself. Meaning, even less so sense, does not therefore reduce to signification; there is no simple one-to-one correspondence of word to thing but rather an unsettled or insecure relation, a type of relay between them. We might call it a dis-*integration* or *disintegralization,* less a decomposition or entropic fragmentation that preempts reference and produces only random sense than a reordering of interior composition vis-à-vis exterior configuration. A thing or an idea cannot enter into a relation with a word on the basis of a presumed conceptual integrity, nor can a word enter into a relation with an idea or a thing on the basis of a presumed verbal and semantic integrity, for that would produce no meaning or sense: we would be (impossibly) back to stones that are only stones, which means at best a nonrelation, at worst a collision.

We will have already seen the effect of exscription operating within Bataille's writing itself, in *Blue of Noon,* in the unsettling effects—which that author grapples with, but also exploits—of language that seeks, as it were, the sovereign moment of its own inadequacy to itself and to the world. That unsettling allows blood, though not explicitly mentioned, to overflow into the "oppositional" soil of a putative revolutionary erotics through reference to trench warfare. Yet the trench is already a figurative "overflow" of the furrow Troppmann has "plowed" in Dirty's body, and any such ex-*humation* or excavation is by definition the soil's overflowing of itself, a hollowing out and overturning of the earth that piles up, in unruly yet regular form, on the edges of what it produces. The possibility of a rhetorical connection between soil and blood thus relies on "exscription" as something like the abyssal, originary, and elemental furrowing of the world by which sense is produced. We would have to read it as the disorderly margin, verge, or shoulder on the edge of that furrow, which interrupts the regular *boustrophedic* lines—back and forth in the manner of an ox pulling a plow—as much as does a writing read from left to right, such as Derrida analyzes in *Of Grammatology*.[8] Though exscription allows for the range of rhetorical operations, such as metaphor, it cannot reduce to any such operation, or even to what we understand as the relation of signifier to signified. The dirt on Dirty's body is no more or less literal than the mud of the trenches; nor is the blood that flowed in those same trenches in a simple metonymic relation to that mud as long as blood is also necessarily implied by any lovemaking, whether that takes place within the cleanest of sheets or in the extremes of perversity. Once sense is made by means of exscription, it is made as a type of pell-mell contagion whose excesses are, at the outside, inseparable from its more linear or ordered operations.

Thanks to exscription, therefore, the stone or the heart comes out of itself sufficiently to mean "stone" or "heart," sufficiently to be represented by those words, and the stone comes out of itself sufficiently to become a heart, a heart of stone, concentrated within itself but "only to the extent that, simultaneously and identically, it is completely exposed outside itself" ("Heart," 171). In exposing itself as heart (of stone), the "previously" immobile stone (at the heart of things) neither remains a *simple* stone nor *simply* transforms itself into a heart. What occurs, however, at the outside limit of itself is something of an *animation* of the stone, the sound or feeling of a beating within the inanimate. And once that occurs, it is clear that the stone that "begins" at the heart of things, but only so as to expose itself outside itself, is neither just a stone nor just any stone but one that necessarily figures an animal body, more precisely, the human animal body. A stone that is a heart, and a heart in general, begins being conceived as and at the bodily center of things, as a corporeal stone, which makes its exscription all the more radical in its consequences for thinking, for the subject, for the body, and for the relation of inanimate to animate—indeed, for life itself.

The obvious model for exscription as sense making is, therefore, corporeal: the body, our own body, provides our primary and perhaps sole basis for understanding how bodies, let's say things, enter into relations; that much seems self-evident. Hence we observe something like the stars as celestial bodies or two objects meeting according to the laws of thermodynamics as bodies possessing a certain mass. But when it comes to a relation that is posited as exposure or exteriorization, one that involves, for example, revealing to the outside different internal layers, or an internal hierarchical ordering, or cooperation and coordination among elements, indeed, any internal organization, which is what happens when sense is made and language does its best to be adequate to the world it is describing and bringing to signification, then the body, with its internal workings and external interactions, presents itself as an even more reliable model for such operations. As a result, sense comes necessarily to exceed or overflow the semantic and to have ascribed to it both the sensorial and the sensual; it comes to be thought of, for Nancy, as tactility, both tact and touching.

The touch of sense as exscription is Nancy's strategy for rethinking the dualisms (e.g., mind–body, but also depth–surface) of Western thinking since classic times, but especially since the modernity inaugurated by Descartes.[9] In his 2000 interview, Nancy uses the word *exogastrulation* to accentuate, as if viscerally, the traditional misconception or misperception of the body as interiority: "The body is the extension through which I touch everything, everything touches me and through that very contact I am separated from everything. The body is what puts me outside, in the sense in which the subject is always outside of itself, it is me as exteriority. . . . The body must be thought of as exteriorization, as a sort of 'exogastrulation' of the subject, a word from biology that defines how certain animals live thanks to an exterior stomach."[10] Such thinking has us imagine in simple causal terms—science typically confirms this—that we touch something outside our body and a sensation travels through the epidermis into our nervous system and brain, where we feel it; we imagine that sort of linear relay of externality to interiority. But it would not be too fanciful to reconceive that operation on the basis of a brain that doesn't just sit passively at the back of our skulls waiting for a sensation to be transmitted to it but that *consists,* rather, in the whole network of transmissions linking skin to cerebral cortex, which means effectively that the brain is permanently exposed or exteriorized. Instead of a relation between two organs, say, skin and brain, we should imagine a network of sensorial interaction.[11] Let us consider what occurs when we don't just touch some thing outside our body but touch some body, somebody, in all the ways that humans do that, from all the coded social forms of greeting and so on, to the effervescence or effusiveness of an animated conversation, to an accidental brushing against in public or private, to a purposeful contact (a hug) with emotional content, to all the touchings impelled by desire that include not touching at all as well as myriad versions of lovemaking, and go all the way to violence and cruelty. We know that as soon as it is a question of those tactilities, we are far from linear models of transmission of sensation, because, in the first place, there is some form of reciprocity, from repulsion to total abandonment, but also because, when we touch, increasingly as we touch, we feel it elsewhere than in the brain, elsewhere than where we touch or are touched—we feel it in our mouths, our stomachs, our entrails, our sex organs, our blood, and our hearts. And we feel those insides outside, exposed, at least to the extent that blood rushes to the surface and to extremities, but more generally because in being touched on the outside, we have also been touched inside, willingly or unwillingly. And the very quandary of what willingly or unwillingly, pleasurably or unpleasurably, means in such cases reposes the whole problematic, for intent and pleasure, presumed to reside somewhere secure inside, are themselves exposed in the same movement.

Those senses of sensing, and of corporality, are touched upon every which way in Nancy's fragmentary essay *Corpus,* from 1992. To recapitulate three moments of that logic, (1) as soon as there is a word, there is overflowing with respect to it, evoking an extension and a touching that is essentially corporeal—"exscription ... detaches words from their senses, always again and again, abandoning them to their extension. A word, so long as it's not absorbed without remainder into a sense, remains essentially extended between other words, stretching to touch them, though not merging with them: and that's language as body."[12] (2) The body is necessarily a relation to exteriority: we don't need someone or something else to touch us to know that; it is brought home even when we touch ourselves, for in doing that we feel ourselves from outside of ourselves—"I have to be in exteriority in order to touch myself. And what I touch remains on the outside. I am exposed to myself touching myself. And therefore . . . the body is always outside, on the outside. It is from the outside" (128–29). (3) As a consequence, the body functions as a generalized *articulationality* or *prostheticity,* which means that from supposed center to outermost extremity, there is an unending economy of negotiation of otherness. The articulation of bones and members taking place within the body is conditioned by the fact of a necessary or primary originary articulation with what is outside the body, in the manner of the pseudopodal syndrome that we have already examined: "Form means that the body is articulated . . . as the relation to something other than itself. The body is a relation to another body—or a relation to itself" (127). In my terms, that has meant, from the beginning, a negotiation with what is foreign: foreign bodies, inanimate bodies, prostheses—an inanimation of the human.[13]

Inside the body as traditionally conceived interiority exposed to touching, it is less Descartes's mind or *ego,* or the brain, that occupies the privileged internal seat than the heart whose beating we constantly feel inside us. The heart is the center of that conception of the body; it goes to the heart of how we understand the body and, indeed, how we traditionally understand centrality vis-à-vis peripherality in general. The functioning of the heart depends on its being a center and a ground for and of the body, which in turn requires for it an unexposed interior security: that of the heart of stone, as we saw earlier. But if such a heart of stone is "concentrated within itself only to the extent that . . . it is completely exposed outside itself" ("Heart," 171), how much more so the heart of love, the heart in love, the heart in the "shattered" state that Nancy defines as "the extreme movement, beyond the self, of a being reaching completion."[14] In French, that definition reads, "l'amour est le mouvement extrême, au-delà de soi, d'un être s'achevant," and we would need to understand the verb *s'achever,* translated as "reaching completion," not just in that sense of accomplishment or wholeness that we feel when we feel in love but also in its more sinister sense of providing a coup de grâce. Love is also the extreme movement outside of ourselves that finishes us off as subjects. Love breaks the heart by definition, but not just by cracking it inside, causing us to sob, palpitate, and gasp for air; it does so more fundamentally by breaking open both the body and self within which it is supposed to be enclosed, exposing our whole being. We cannot remain ourselves and be in love, or we cannot be ourselves and remain in love. Before getting to a point of accomplishment where we feel that our exposure is healed and safeguarded by the reciprocal exposure of another, indeed, even when we again feel self-enclosed as two hearts are supposedly joined as one, love has nevertheless exposed us, and our selves, to alterity itself. It exscribes our self, our being, and finally our life as the *prosthetic articulationality* that is a constant theme throughout this study.

Given that, it is hard to imagine how one can be in love, how one can *be* in love where love means that sort of fracturing of being, at least of *being subjective*; and conversely, how love can be, when *being more subjective than ever* is precisely what we expect from it. In the first place, outside of or before everything that has been developed up to this point, we know in fact, in spite of what we expect and even through those expectations, that being in love produces a type of subjectivity that, for being heightened in many respects, is profoundly disorienting vis-à-vis our sense of being. The overwhelming feeling of well-being that love induces is nevertheless disturbing; its strong sense of purpose is nevertheless accompanied by an extraordinary passivity that begins with the inescapable impersonal force of *falling* in love, something that necessarily happens in spite of ourselves. And our energized and racing heart on one hand leaves us light-headed and on the other impels us toward a climax that is one form or another of utter ecstatic abandon and, finally, involuntary spasm.

Perhaps I am again confusing love with sex. But we seem to be inexorably drawn to this conclusion: if we really want to know what love is, we must, paraphrasing Nancy's formula for writing, touch its extremity, its ecstatic moment.[15] That presupposes, of course, that we know what sort of love we are talking about, which is a question that I have been begging all the way through this chapter, failing to satisfactorily distinguish among different forms of it that Nancy considers inextricably mixed: "charity and pleasure, emotion and pornography, the neighbor and the infant, the love of others and the love of God, fraternal love and the love of art, the kiss, passion, friendship" ("Love," 246). It presumes that we will have bracketed the whole history of love—even if we were to restrict that to the Western world—through mythology, romance, chivalry, courtly love, religious passion, romanticism, and so on, as in Denis de Rougemont's classic study, and that we will have distinguished that, for example, from a history of sexuality such as Foucault's or a sociology such as Luhmann's. It would also require us to keep in reserve a number of contemporary discourses other than Nancy's, such as that of Marion.[16]

It would mean concentrating on physical love, presuming we knew where that begins and ends, and it would mean distilling that into a more or less purely physiological moment, that of the orgasm. But if love represents the specific instance of corporeal exteriorization that I have been describing up to this point, then it seems feasible to concentrate on its extreme bodily moment as the privileged figure for its operations in general: an exploration of body surface and cavities constitutes physical love in the obvious sense, but more specifically, the lover is prey to a more or less unmanageable externalization of the blood, beginning with blushing, flushing, and blotching of the surface of the skin, as well as swelling, engorging, and various tumescences, ending with the contractions of orgasm itself. In that climax, it is as if the sexual organs had taken over, in the very moment where the heart skips a number of beats, the heart's repetitive life-affirming function; it is as if the sex organs were beating like that heart. Sexual climax would precisely be, in that sense, a type of externalization of the heart: aerobic exertion; blood pulsating on the body's surface and extremities; fluid excess and overflow; life on the edge, in the extreme, assenting to life all the way to death as Bataille formulated it.[17]

In the extreme, therefore, love overflows outside life. Nancy is explicit in describing the joyous *alter*-ation or exposure to the other that takes place when we come, and which means, to repeat, not just receiving the otherness that comes from another but, more importantly, subjecting one's self to its own otherness, an otherness that we would have to understand as including something other than human animateness. At this peak of life, there emerges something other than life as we presume it to be: "Coming [*jouir*] means being traversed by the other. The other cuts across me, I cut across it. Each one is the other for the other—but also for the self. In this sense, one comes in the other for the self. . . . This is the syncope of identity in singularity. A syncope: the step marked, in a suspense, from the other to me, neither confusion nor fading, clarity itself, the beating of the heart, the cadence and the cut of another heart within it" ("Love," 271, translation modified). A syncope occurs at the (outside) heart of the heart—once it is exposed or exscribed in love, by love—a syncope that denies the heart its cardiac essence, or at least its essential rhythmic certainty. The heart breaks itself by breaking its own rhythm, by syncopating itself, cutting a piece out of its beating. *Syncope* literally and etymologically means the cutting from within of a togetherness. In rhetorical terms, it is the ablation of a letter or syllable from within a word, such as when "forecastle" is pronounced, then spelled, "focsle," and in medical terms, it refers to a swooning or light-headedness caused by an irregularly pumping heart and deficiency of blood in the brain; and of course, in musical terms, it means a change in rhythm, the accenting of a traditionally unaccented beat. The heart beats, and *is,* according to Nancy's definition, as that very syncopation of itself: "Actually, the heart is not broken. . . . It is the break [*brisure*] itself that makes the heart. . . . The beating of the heart—rhythm of the partition of being, syncope of the sharing of singularity—cuts across presence, life, consciousness" (263).

It is in that most organic of organs that corporeal exteriorization begins, or is constituted. The shattered heart that allows us to be in love disturbs not only our being-self-enclosed but also our being-alive. At the core of life, in the beating heart conceived of as compact interiority within the interiority of the body, there is an interruption, an irregularity, a suspension of flow that threatens consciousness and therefore life. The heart skips a beat; there isn't enough blood. Such a spasmodic interruption is easily understood as the heart's outside possibility, clearly recognized at the fullest of life, in the sexual ecstasy where heart and body burst and overflow, in the throes of an uncontrollable orgasmic convulsion that shakes the body to its core like some cruelly fulfilling mimicry of the beating heart transformed into pure irregularity, stopping the blood in fits and starts of syncopated flux. But if the orgasm is where we register the explicit effect of an ecstatic body bent on blocking its natural flow, we would have to understand a certain structure of interrupted flux, of *bloodlessness* finally, to inhabit and even constitute the beating heart wherever it is found, either safe in some interior repository or broken open by the force of love; we would have to understand interruption as the very constitutive possibility of a heart whose regular beating in fact operates as a spasmodic interruption of itself, alternating contraction and dilatation, systole and diastole.[18] The body beats outside in fits and starts when we come, as though its heart were exposed as pure bloodless beating, but such an exposition is the heart's innermost possibility, that of being at one and the same time inside pulse and outside blow. *Un coup* in French means that blow, cut, or shock, either subtle or violent, but it could as well describe the throb of a heartbeat. In the expression *un coup de coeur,* it refers to a sudden attraction, one more likely leading to an impulsive purchase than anything else (when one falls head over heels in love, one is said rather to have been struck by lightning, *un coup de foudre*). But *un coup de coeur* suggests nevertheless that the heart is turned away from its primary purpose, from its regular beating, and exposed instead to a shock that interrupts it. And it also suggests that the heart is, in the gentle shock of its regular beating, in the very series of *coups* that constitute its primary purpose of assuring the uninterrupted flow of blood, diverted from that purpose. Every beat of the heart functions also as an offbeat; when it comes to the heart, every *coup* is, at the outside, bloodless, which does not mean for all that without violence.

Outside, bloodless—would we also have to call such a heart lifeless? For whether it be operating deep inside, in the simple repetitive reaffirmation of itself, repeating the vital impulse via automatic electrical firings of the sinoatrial node to make the heart beat some 2.5 billion times during an average human life-span, or in the outside fulfillment of itself, in the ecstatic affirmation designed as triumph and perpetuation of life, in the quickening expansiveness of unbridled passion; whether in one or the other place, or on one or the other side of the coin of life as we know it, define it, or feel it, there is an interruption of life. Life is experienced in both "places" as an interruption of itself. Straightforward binary electronic logic tells us that for every instance of "on," there is an instance of "off," experienced also as the risk of overload or simple failure; and the climactic apex that accomplishes life cannot be constituted without a falling off from that summit of paroxysm, passion, and joy; orgasm would not be a peak of pleasure and life that we had reached if the following moment, just like the preceding moment, were not something less elevated. Love teaches us that life is affirmed, at one or the other extreme, by interrupting itself.

The bloodless heart, exposed in love, out here, beating, or rather *offbeating* outside of itself, or beside itself with amorous distraction, scans also the stark repetitiveness of its beating and brings us back once again to the fact of mechanical automatism. The body is never more aware of the oxymoron of its organic mechanism than in the case of the beating heart, in the automatic electric or electronic pumping operation of the organism that is one basis for determining at least vertebrate life—the organism lives thanks to oxygen pumped in blood by the heart to the brain—but whose functioning is also characteristic of the lifeless or inanimate machine. The exscribed heart is in the grip *(sous le coup)* of technology, beating outside the body to the extent, finally, of being a transplanted and even artificial heart pumping synthetic blood, possibilities or actualities that Nancy recounts or gestures toward in discussing his own heart transplant in "The Intruder."[19] Such a heart is defined by a type of technological bloodlessness that we have nevertheless increasingly to recognize as a form of life, an inanimation that comes about not because of a past, present, or future innovation—such as a prosthetic electronic heart—but rather because the heart, like the body in general, was prosthetized as soon as it opened itself to some form of relation or articulation to the outside, to any foreign body whatsoever, which means from the beginning, and was automatized as soon as it began beating and pausing from beating, which means always. We would have to understand the quickening or animation of the heart that takes place in love not just as its yearning for, or movement toward, another such heart but as an impulsion toward otherness itself and in general, including the inanimate; and we should understand it also as the heart's own inanimate drive, beginning with its syncopation and leading to its utter exposure, to its being shattered, bled dry. Life no less than love, I am insisting, can be said to begin in and as that syncope; that is its irregular kick start. And if our lifeblood rushes to epidermal extremities and tumescent organs in the flush and tussle of love, it would be not because the heart safe inside has pumped it there but rather because the blood is rushing after the heart that has already left for life on the wild outside.

The heart beats, the telephone rings; absolutely no connection?

Syncope has its opposites, or complements. The first is "apheresis," which means, to quote Thomas Wilson in 1553, for example, "Apheresis. Of abstraction from the first, thus. As I romed al alone, I ganne to thynke of matters great. In which sentence, (ganne) is used, for beganne." In medicine, apheresis is a case of hemapheresis, referring to the circulation of blood outside the body, through an apparatus that separates its components (e.g., platelets, plasma) before returning it—minus the removed component—to the veins. I'll allow that idea of bloodletting, or structural bloodlessness, that of a prosthetic supplement, to circulate back through all that has just been argued.

Syncope's second differential (oppositional, complementary) form is "apocope." Wilson again: "Apocope. Cuttyng from the end. A faire may, for maide." One can and may, verbally, even according to Webster, apocopate as well as syncopate. In medical parlance, *apocope* is synonymous with *abscission, ablation,* or *amputation.* Indeed the Greek preposition *apo-*, like the Latin *ab-*, "away (from)," makes *apocope* signify the general category of cutting away or removal, of which syncope would be a specific case. Or perhaps, conceptually speaking, syncope is rather the general category of interruption or rupture that is a condition of possibility for a more extreme apocope. Those ideas should also be allowed to beat or skip back through all that has just been argued to emphasize the structural necessity that ties even the slightest rupture—in turn, but not in a single direction—to removal and prosthetic replacement. The possibility of prosthetic replacement, as I have argued here and elsewhere, is what in turn, but as it were in reverse, posits or conditions the possibility of amputation; amputation is structurally inseparable from, and informed by, the perspective of prosthetic replacement. By extension, as I am reinforcing here, prosthetic supplementation is posited by syncopation in general. The wager here would be to conceive of a heart that skips a beat *because* it can be replaced, by that of a donor, or even by an artificial one with continuous flow—to conceive, as it were, of a natural heart that envisages an artificial one.[20]

Every interruption carries with it the idea of an absolute rupture, every heartbeat the syncopal possibility that it will be the last. The heart beats in the prospect or expectation of ceasing to beat; every beat, as we have already insisted, is also a syncopal cessation of beating. With every interruption, as with every heartbeat, there is a caesura, a pause and a waiting, with the expectation, but never the absolute certainty, of a resumption. Behind every beat, as if shadowing it while constituting it, is the absence of a beat. Absence, *apousia*, of course describes originary rupture in general, and waiting in the absence of the loved one would be the *apocope* that constitutes love, but only—according to the logic just developed—once waiting has itself first been constituted by the prosthetic supplement of something that will repair its lack. In Roland Barthes's analysis of the condition of loving, that prosthetic supplement is consistently the telephone. It seems that the telephone is the condition of possibility of waiting, which is the condition of possibility of absence, which is, for Barthes, the condition of possibility of love: "The lover's fatal identity is precisely: *I am the one who waits.*"[21]

Waiting occurs in the space of an interruption, in the space of a temporal rupture. One waits within the general structure of what could be called telegrammatic distancing, a spacing or delay over which communication is both necessary and impossible. One waits, in that sense, for what one doesn't know to be coming, not with absolute assurance. Waiting with the assurance of an arrival wouldn't be true waiting, which is precisely what is highlighted by the waiting of love. Every waiting carries with it the fear or threat that whomever or whatever one waits for won't arrive; every waiting lover knows that fear and threat. Within that general structure of the *telegrammatic,* the telephone inaugurates, in the first place, the contemporary phase of presencing whereby the voice provides instantaneous, if not immediate or unmediated, contact across an irreducible absence. Different from telegrammatic communications such as letter writing, e-mail, and texting, where a type of touching occurs by means of the transfer of handwriting to reader—"the page itself is a touching (of my hand while it writes, and your hands while they hold the book). . . . In the end, here and now, your own gaze touches the same traces of characters as mine, and you read me, and I write you," writes Nancy (*Corpus,* 51)—the telephone (like subsequent forms such as Skype) offers sensorial and sensual contact, of the mouth and in the instant.[22] Derrida, in *On Touching,* thus evokes separated lovers who carry on their relationship by means of "telephonic memory":

> It would be time to speak of the voice that touches—always at a distance, like the eye—and the telephonic caress, if not the (striking) telephone [*coup de téléphone*].
>
> (Imagine: lovers separated for life. . . . On the phone, through their vocal

inflections, timbre, and accent, through elevations and interruptions in the breathing, across moments of silence, they foster all the differences necessary to arouse sight, touch, even smell, so many caresses, to reach the ecstatic climax from which they are forever weaned—but of which they are never deprived. . . . They have faith in the telephonic memory of a touch . . .)[23]

In the second place, the telephone defines the general structure of a surprise, or what I have elsewhere described as a dorsal interpellation. However close to it one waits and however much one expects a call, its ringing has the capacity to startle like Althusser's policeman's whistle.[24] Its electronic automatism is mirrored by the body's adrenal automatism in a type of sympathetic cooperation: not just that when it rings one jumps, even and especially when one is waiting for it to ring, but also that the expectation of its ringing speeds the heart as though it were already ringing. It may not differ in that respect from other such (un)expected sounds: the click of a gate, or lifting of a mail flap, or the doorbell itself that announces a lover. But the telephone has a specific prosthetic relation to the body, disturbing its boundaries. And even though it does that, logically speaking, from outside, adding a supplementary boundary, it also produces something of an intrusion within the body. That is so by virtue, most obviously, of being held to or inserted in the ear, but more generally in its requiring passivity and proximity, staying close, reconfiguring corporeal economies such as eating and excreting. Barthes describes the strictures of a pre–cell phone regime as follows: "I have received *orders not to move.* Waiting for a telephone call is thereby woven out of tiny interdictions, *ad infinitum,* even unavowable ones: I forbid myself to leave the room, to go to the toilet, even to telephone (to keep the line from being busy). . . . The anxiety of waiting, in its pure state requires that I be sitting in a chair within reach of the telephone, without doing anything" (*Lover's Discourse,* 39). And in the case of the portable phone, of course, the apparatus is henceforth literally worn as a prosthesis; in the prime example, it sits in the breast pocket, ready to vibrate or ring like an electronic heart on the outside of the heart.

The sound of a telephone ringing has a specific phenomenology and semiology, akin to those of a voice, and thanks no doubt to the telephone's metonymic and prosthetic relation to the voice. It calls as much as rings, sounding differently, first, according to variations in distance, clarity, and echo. But it also calls, by virtue of its surprise, through the differences of a more or less unconscious space and medium, doing so more literally when it sounds through the night and through the fog of sleep (or stridently cuts across the chimes of domestic or municipal clocks and parish churches). Thus the phenomenology and semiology of the telephone is inextricably linked to a geography and sociology, relating to the specifics of the space, both interior and exterior, in which it rings; and now to an *informatology*, an account of its limitlessly extended range of ring tones. However it sounds, though, the ringing of a telephone operates through the space of a syncope: like the heartbeat, its discontinuous sound involves both interruption and waiting. Its ringing is perhaps more like the heartbeat in the double-tone spondee chime of the United Kingdom and certain ex-colonies than in the single prolonged ring that until recently dominated landlines in Continental Europe and North America. Since childhood, I will have never been able to escape the haunting of a sentiment that in the caesura between the two rings, or two parts of a single ring, death falls. Especially in the dead of night. Not just that the call that arrives in the dead of night has as its structure the announcement of a death (or some such unexpected accident for which death is the paradigm) but that the ring itself, in interrupting itself, necessarily announces death in the form of its own syncope and desistance. Never escaped that since the call my father picked up, hurrying with his very own syncopated footsteps to reach the phone too late before it woke the whole house, to be told his younger brother had flown his plane crashing into telephone wires, electronic caesura come full circle. Or perhaps, more simply, never since hearing the melancholic spondaic night song of the onomatopoeic morepork, ruru, or *Ninox novaeseelandiae,* calling without waiting for an answer, or else ending in a screech, like some adestined telephonic alarm bespeaking abandoned love or impending death.[25]

The lover's fatal identity is precisely that of one who waits: by the telephone, *like* a telephone. "Like" here means operating through a similarly syncopal structure, involving interruption and therefore waiting: not just that the lover waits, with syncopated heart, for the telephone to ring but that the lover waits for the telephone as the telephone waits, through its own interruption, for itself, for the sound of itself, and that the lover's body is both prosthetically connected to an exscripted heart that beats syncopatedly and tied in obedience to a telephone that rings syncopatedly (un)expectedly.

In *A Lover's Discourse,* on first reading, the lover waits in an implacable state of mourning—understood in classical psychoanalytic terms—for the lost object ("I am an amputee who still feels pain in his missing leg" [39]), for the mother's breast, or for the Mother. The prosthetic status of the telephone within that economy derives from its being an instrumental machine capable of relaying the lover (back) into contact with the lost object; when it rings, the lover will have received "the healing call, the return of the Mother" (38–39). The telephone serves thus as an almost fetishistic focus for various neurotic twists and turns, the point of triage for aspects of the lover's condition: the subjugation or physical passivity already mentioned (82), a concomitant defamiliarization of other objects (87), the threat of suicide (178), the paranoia of oversignification (63). The success of mourning is determined by the possibility of a detachment that allows the telephone to resume "its trivial existence" (107), by being deactivated from that mechanism of neurotic relay and fetishistic focus.

Once love is mourned, then, a telephone can return to being just a telephone. But as long as love is love, active and mobilized, so too is the telephone, and in such a way that another reading of it becomes necessary, where its function does not reduce to a device for communicating lack. For the amorous relation is constantly haunted by a series of non-instrumental *telephonic* effects, precisely effects of the voice in relation to distancing. What love mobilizes is simultaneously telephonic memory and the threat of telephonic amnesia, in other words, a voice distanced by absence and forgetting:

> The voice supports, evinces, and so to speak performs the disappearance of the loved being, for it is characteristic of the voice to die. What constitutes the voice is what, within it, lacerates me by dint of having to die, as if it were at once and never could be anything but a memory. This phantom being of the voice is inflection. It is inflection, by which every voice is defined, that is in the process of falling silent; it is that sonorous

grain which disintegrates and disappears. I never know the loved being's voice except when it is dead, remembered, recalled inside my head, way past the ear.[26]

Given that idea of the voice, more particularly the inflection or grain of the voice—the "phantom being" of its inflection—as paradigm of the loved one's mortality, that loved one will die first on the telephone, in a sense being put to death by the telephone, placed by it "in a situation of departure":

> The loved, exhausted heard over the telephone is the fade-out in all its anxiety. First of all, this voice, when it reaches me, when it is here, while it (with great difficulty) survives, is a voice I never entirely recognize. . . . Then, too, on the telephone the other is always in a situation of departure; the other departs twice over, by voice and by silence: whose turn is it to speak? We fall silent in unison. . . . *I'm going to leave you,* the voice on the telephone says with each second. (115)

It is precisely the telephony of the voice that, by inscribing distance, determines the voice's disappearance and sets in motion the work of mourning. Because "the telephone wire is charged with a meaning, which is not that of junction but that of distance" (115), the voice of the lover heard on the telephone produces the whole anxious complex of loss: fade-out due to that distance, fade-out due to exhaustion, double fade-out of departure. According to the classic reading, on one hand, the telephone is the bad object that installs distance rather than nullifies separation.

On the other hand, what we have just recognized as a type of *telephonization* of the voice, the fade-out and departure of its grain, begins simultaneously with love itself. For the mourning that defines love as waiting is said to be not contingent but originary—"the lover's anxiety . . . is the fear of a mourning which has already occurred, at the very origin of love" (30). Simply put, love, no more than desire in general, does not emerge from the psyche and set about finding an object. The object, recognized as other, renders love possible at the same time as it renders it finite. Such a logic constitutes love itself as originarily telephonic. Just as the loved one will have always already left, just as that departure will have always already been announced by the phantom being that is the inflection of the voice, so that fade-out will have always already been heard *on* the telephone and *in* the telephone. That is to say, first, that one does not require a telephone as apparatus to hear the loss of inflection; it is heard as a natural function in the case of fatigue: "Nothing more lacerating than a voice at once beloved and exhausted: a broken, rarefied, bloodless voice ... such a voice is *about to* vanish, as the exhausted being is *about to* die" (114). But second, it means not only that the telephone amplifies the voice's natural loss of inflection caused by fatigue but also that every voice is a fatigued telephonic voice: every voice is endowed with an idiosyncratic inflection, doubled by another tone, by a phantom that is capable of leaving it, that is in a sense already dead within it, that is the death it carries within it. Every voice has an inflection that is distanced from it even as it comes back to it; every voice has a ringtone that is always more or less on or off, like a telephone that rings and so heals love, or doesn't ring and so puts love to death. Every voice has a voice on the telephone, saying at every second *I am going to leave you.*

The other's fatigue, heard as dying inflection of the voice, its departing phantom being, is said to "alter" the imaginary fullness of love, to fracture it and render it infirm. *Une voix altérée* is precisely a broken voice, one wracked by fatigue or grief. "Alteration" is one of the figures of the lover's discourse whereby "according to minor incidents or tenuous features, the subject suddenly sees the good Image altered and overturned" (25). It is also described as a "counter-rhythm: *something like a syncope* in the lovely phrase of the loved being" (25, emphasis added), and it occurs most commonly in the language used by the loved one, in her borrowing a word or phrase from an alien world or register ("I hear rumbling menacingly *a whole other world,* which is the world of the other" [26]), something foreign to the "cocoon of my own discourse" (27). Barthes gives the example, from Proust, of Albertine's recourse to the expression "get her pot broken [*se faire casser le pot*]," which suddenly reveals to the narrator "the dreaded ghetto of female homosexuality, of crude cruising" (26), which could have us understand "alteration" as a type of queering of language.[27] But we should perhaps hear altering as signifying, more literally, the general case of language's departure from itself in the manner of an inflection of the voice, the general *telegrammatization* or *telephonization* of language that means that it necessarily comes from elsewhere, from a foreign place, from the other and from otherness: "the other ... lets be heard, by a word escaping unchecked from his or her lips, the languages which can be borrowed, and which consequently others have lent" (27). An "othered" voice *(une voix altérée)* can, in its brokenness, be received at the same time as the distillation of an idiosyncratic inflection, as the lover's purest *ownmost* idiom or most germane grain, and as the very failure of that particularity and the failure of the voice in general: telephonic breakdown, no dial tone. Similarly, language, by virtue of its pluralistic idiomatic structure, by virtue of its being able to depart from any imagined purity of communication—which is precisely what allows the particularities of individual expression—also necessarily falls, out of itself, *syncopally,* into otherness, into the "alterations" of what is borrowed, colloquial, or crude. The language of love, as much as any other, is required to negotiate those ruptures. To write (of) love, for example, is to deal with "the illusion of expressivity" (98); indeed, it necessitates renouncing a utopian language of the imaginary in favor of the disorderly and wasteful: "To write love is to confront the muck [*gâchis*] of language" (99). More specifically, the love letter, "purely *expressive,*" fails as communication because of its dependence on a reply: "without a reply the other's image changes [*s'altère*], becomes other" (158); the amorous correspondent is as a result exposed to "the 'injustices' of communication" (159). Communication is even more illusory in the case of the argument or dispute that Barthes refers to as the "scene," where "language begins its long career as an agitated, useless thing" (204) but that nevertheless can continue interminably. The scene "is that language whose object is lost" (205), divorced from its referential function and become in that sense completely *altered,* diverted, or some would say perverted from its purpose, but by the same token limitlessly expansive within its own narcissistic space: "the scene is interminable, like language itself: it is language itself, captured in its infinity" (207).

The language of the scene takes leave of itself; in becoming pure rant, uttering no more than itself, it cracks; it reduces to either pure intonation or nothing recognizable, all telephone. One would expect to find its opposite figure in the declaration, or as Barthes calls it, the exclaiming or blurting out *(profération),* of love, its repeated cry, preferably a single agglutinated word (as in Hungarian): "I-love-you" (147). However, in his analysis, "I-love-you" is found to function as a more or less pure performative, emptied of constative content in the same way that the language of the scene was emptied of its object. Like any performative, it cannot function if uttered one time only but derives its "authority" precisely from being a formula that is enacted over and over. There where love should signify to its utmost extent, where the loved one should be told, let's say, the truth, in no uncertain terms, he is indeed told, but is told nothing, no thing: "Once the first avowal has been made, 'I love you' no longer means anything.... There is no other information in it but its immediate saying" (147, 148–49). To its credit, the performativity of I-love-you means that it approaches music (149), affirms like an "amen" at the limit of language (153), and celebrates, by imitation, the expenditure of lyric poets, liars, and wanderers (154).[28] But each of those effects takes it further away from its communicative and informative purpose, once again as if stripping language away from itself, like an inflection, structured by its mortality: language as telephonic syncopation or syncopal telephony.

What distinguishes I-love-you, in Barthes's terms, from those other performative forms—music, affirmation, expenditure—is the fact of its requiring a response, without for all that being a question. In that, it is like the love letter, but whereas various responses are possible—the only unacceptable one being "there is no response" (149)—the desired ideal response remains out of reach: the ideal symmetrical and reciprocal response to I-love-you, the chiasmus of the same thing repeated back by the loved one, is something that language seems not to permit. "I-love-you" is not a response to "I-love-you," which instead requires some disequilibrium, however slight, such as "I love you too" or "so do I" *(moi aussi)*. Neither of those is perfect, for their form is "deficient, in that it does not literally take up [*reprend*] the proffering" (150), in the sense of repeating it. One would have instead to envisage the empirical possibility of "our two profferings be[ing] made *at the same time*" (150), which would have the converse effect of annulling the economy of exchange that we presume to be necessary to the love relation, even as we aspire to surpass it.

The simultaneous double proffering of I-love-you, supposing it were possible, would complement the passionate embrace and union of two bodies or the sympathetic beating of two hearts as one. It would say what we mean to say by love, defined as something that doesn't need to be said: love that expresses itself while remaining contained within its interior security, as sure as a beating heart. But unless it is to revert to stone, love cannot remain within its own secure confines. That is true for any type of love, and for the act as well as the word of love. The truism is true: love begins by reaching out, which means that the word of love is exscription par excellence. The single necessary word that, it seems, must be spoken if there is to be love, the word for which there is no substitute, inscribes love on the outside of sense far from the safety of the heart that gives rise to it, at best in the *contentlessness* of a universal formula, at worst in the crassness of the most hackneyed, "the most worn-down of stereotypes [*du plus éculé des stéréotypes*]" (151).

Perhaps an inflection can save I-love-you or breathe fresh life into it. Perhaps that is what allows it to be infinitely repeatable, and perhaps it is in the absence of such inflection that the death of love begins to transpire, that love begins to fatigue and expire. For by requiring that supplementary tonic layer upon the amorous proffer in order to authenticate itself, love will have opened a fracture within itself and announced its mortality from the beginning. In being inflected, the language of love will be subject to degrees of clitic dependence; it will come to be heard as composed, composite, but no longer intact. Love itself will have borrowed the structure of the loved one's voice, subject to fatigue, alteration, and death, inscribed with its own telephony, with at one and the same time effects of intimacy and proximity, of separation and distance.

I-love-you is therefore the heart of love, the necessary discursive essence of love that is proffered or telephonically exscripted on its outside edge. It is heard repeating itself there. Failing a satisfactorily symmetrical response, failing a unison that could in any case only be heard as miscommunication, neutralization, or contradiction, repetition is its single format. Love speaks in that place and in that way with its own syncopal rhythm of measure and caesura, sounding, pausing, sounding again, with each pause threatening to sound no more, a machinic repetition that is both affirmation and autoaffection, an utterance that echoes back to touch itself on the outside of itself. As we understood from Nancy, the exscripted language of love or word of love signifies with a corporeal consistency that is both natural and prosthetic, both an extension to the edge of itself and an addition functioning outside itself, on the border where animate articulates and confounds with inanimate. I-love-you is spoken in the prospect of a copula, it is predicated on a chiastic fusion; but I-love-you both exposes the subject who utters it to that subject's own *disintegr(aliz) ation* and names an object of love that, for seeming to be identified, is far from being defined. The articulation that love operates between "I" and "you" is a disjunction that destabilizes both subject and object, not just in their positions but also in their definitions. In loving outside of myself, I expose myself to every other body, beginning with my own and ending wherever an end to bodies might be found, human bodies, animal bodies, vegetal bodies, inanimate bodies. Love, once articulated, is necessarily *altered* to that extent. It is articulated with a detachable inflection that will either be reattached as part of a new, contrived wholeness or fall outside as a foreign object that the body no longer recognizes, utterly other.

The exposure of love means, therefore, that when I-love-you is spoken, there is an inflected something that either is given and received as a gentle caress or departs and arrives "full in the face [*de plein fouet*]"; either way, it is something supposedly internal to the body that is turned outside to act back on the body. Seeking a symmetry and reciprocity, it finds only repetition, that of its own automated activation. It keeps on going anyway, for that is its mode, enacted as love that has no choice but love like life that has no choice but life, it sets in motion its own rhythm to produce itself as love, to make itself as love, love-speaking indistinguishable from love-making, for, as Barthes says, "what matters is the physical, bodily, labial proffering of the word: open your lips and let it out (be obscene)" (152), say it and do it over and over, I-love-you-I-love-you, or fuck-me-do-it-more, a single chant that loves to hear itself cry it out, then wait to hear it come back from the other or from the same, proffer then pause, proffer then pause, the systole and diastole of the utterance and performance of love, sound and pause or pump and pause, a beating heart or ringing telephone finally, each *coup* equally bloodless, a telephone ringing in the body or a heart beating outside it, beating and ringing to the end of beating and ringing, don't answer, don't stop, let it ring, don't answer, don't stop, let live, don't answer, don't stop, let love.

8

THE AUDIBLE LIFE
OF THE IMAGE

Godard

Since at least 1980, Godard's cinema has been explicitly looking for (its) music, as if for its outside. In *Sauve qui peut (la vie) (Every Man for Himself)*, the protagonist Paul Godard hears, and asks about it coming through the hotel room wall, and it follows him down to the lobby, but despite his questions, it remains "off," outside the field of vision, in the same space as Marguerite Duras's voice, until the final sequence. At that moment, at the end of the section precisely entitled "Music," the protagonist is at the same time struck by a car and struck by the entrance of music into the diegetic present of the film, as the camera pans past an orchestra playing on the sidewalk while Paul fades out of the world under the quizzical gaze of his daughter.[1] By 2004, with *Notre musique* (Our music), music would seem to have taken over the whole text, for the film was announced as being about the collaboration between Godard and German record label ECM. In the context of that film, it is difficult to determine both what that *music* is and who *we* are, although this chapter will try to advance a hypothesis in that regard. In fact, my main contention will be that music in Godard's films functions as something like the *absent image(s)*, not those it has lost but rather its cinema to come, what remains to be discovered and live within it, the future survival of it. Not a cinema that cannot be seen but rather the image that can perhaps only be heard; and not the romantic or psychedelic dream of a synesthetic apotheosis either, rather the technological coincidence of *sonimage* that has also been the precise direction of Godard's cinematic research for about forty years. I shall argue that his music constitutes a further case of technological life beyond the simplistic opposition of animate and inanimate, as if music were an image moving beyond cinema into a form of autokinetic ipseity.

In one respect, already suggested by the name of Godard's production company (Sonimage), but reinforced by a variety of commentators, having music seen, or having film function beyond the formal constraints of image and sound tracks, is what Godard has always aimed for. The filmmaker has himself expressed it in these terms:

> Musicians don't need the image whereas people who make images need music. I've always wanted to do a pan or tracking shot during a war scene or love scene, so that one can see the orchestra at the same time. So that the music takes over when there is no more need to see the image, so that music expresses something else. *What interests me is to see music, to try to see what is heard and to hear what is seen.*[2]

In more general terms, however, in the context of "Godard's life-long innovations in sound design,"[3] one could cite, from as early as *A bout de souffle (Breathless),* the opening of the invisible wall between on-screen actor and spectator; or the whole complex play of sound that begins when Belmondo shoots the sky. As a means of "anchoring" some of those considerations in a musical context, I shall concentrate in this chapter first on two sequences from *Week-end,* where what becomes more explicit in *Every Man for Himself* is already heralded, then develop my thesis further vis-à-vis the relatively recent films *Éloge de l'amour (In Praise of Love)* and *Notre musique* to show how Godard's idea of music is inextricably related to his radical conception of montage, to montage as a radical concept.

My reference to the concept, in the context of cinema, is of course a reference to Deleuze's idea of what philosophy does best: "philosophy is the art of forming, inventing, and fabricating concepts."[4] And any discussion of cinema in the context of philosophy henceforth comes across Deleuze before it is able to advance at all. In the first of his two volumes on cinema, Deleuze returns to the terms of reference discussed in the introduction, describing "a world of universal variation, of universal undulation, universal rippling [*clapotement*] . . . a kind of plane of immanence" (*Cinema 1,* 58–59). Here, however, the relations or events that function across the plane are identified, following Bergson, specifically as images, and even more specifically as images defined by their very movement, as "movement-images": "The material universe, the plane of immanence, is the *machine assemblage of movement-images*" (59). It is on that basis that Deleuze will advance—again following Bergson, but also in spite of that philosopher, for whom cinema simply repeated the limitations of natural perception—the concept of a cinematic perception, as it were, not *of* the world but occurring *as* the world, a concept of "the universe as cinema in itself, a metacinema" (59). Cinema for Deleuze is far from a medium that allows the spectator to better perceive the world; it is rather something that "thinks" like the world in its insistence on a perceptibility that, for being all movement, has very little to do with the natural perception of objects obeying a Newtonian-type billiard ball model. Thus, in a passage from the "Becoming" chapter of *A Thousand Plateaus*, it is precisely imperceptibility that is related in turn to movement and to cinema:

> Movement has an essential relation to the imperceptible; it is by nature imperceptible. Perception can grasp movement only as the displacement of a moving body or the development of a form. Movements, becomings, in other words pure relations of speed and slowness, pure affects, are below and above the threshold of perception.... What we must do is reach the ... cinematic threshold. (*Plateaus*, 281)

To reach the cinematic threshold is in a sense to become cinema, which means developing a whole other relation to perceptibility. By starting from movement, cinema has the capacity to mobilize perception into or toward the impersonal percepts and affects that operate on the plane of immanence, no longer as objects of sensorial prehension or conscious reflection but as functions of the composition of the plane itself: "Perception will no longer reside in the relation between a subject and an object, but rather in the movement serving as the limit of that relation.... Perception will confront its own limit" (*Plateaus*, 282).[5] And similarly, the image, no longer a "static" moving object on a screen, will instead be able to connect the spectator to—as it were, plug in to—the machinic assemblage of everything that moves as it moves, to the motion of universal rippling. Even the action-images of a thriller that appear to engage only the sensory-motor functions—being propelled, anticipating, fearing—will in effect constitute a first step on the way toward that connection; and that will no doubt be much more the case when it comes to the time-images of a "so-called modern cinema," in which movement "subordinates itself to time," becoming "false" or "aberrant" movement (*Cinema 2*, 271). Such a cinema may not manage to give us films that show the infinite images of the plane of immanence, but it can offer "pure optical and sound situations, in which the character does not know how to respond, abandoned spaces in which he ceases to experience and to act so that he enters into flight, goes on a trip, comes and goes, vaguely indifferent to what happens to him, undecided as to what is to be done" (272). As a result, the spectator will be less interested in what is to be seen next than in dealing with the problem of what there is to see, or even the question of what constitutes seeing once it is no longer a matter of a subject seeing an object, rather of something like seeing seeing, hearing hearing—and indeed, for this will be our question—seeing hearing.

The disruption of the movement-image by the time-image also amounts to a virtual doubling of the image: "For the time-image to be born . . . the actual image must enter into relation with its *own* virtual image as such. . . . An image which is double-sided, mutual, both actual and virtual must be constituted . . . *indiscernibility of the two*, a perpetual exchange" (*Cinema 2*, 273). Deleuze describes two main protoypes of such a relation of virtual and actual, namely, memory or dream images, but the more general model for it is what he calls the crystal-image, where, again inspired by Bergson, he describes a breakdown or scission of time that is not just a reconfiguration of present with respect to past but what he calls "time itself [*en personne*], the gushing-forth [*jaillissement*] of time" (*Cinema 2*, 82).[6] As in the case of perception, it is as if access were granted—by no one to no one—to the borders of the total image-world: "the crystal-image [is] also the outer-most, variable and reshapable envelope, at the edges of the world, beyond even the movements of the world" (80–81, translation modified[7]). The time of such a world is of course nonchronological, and its life, as Deleuze writes in one of the rare such formulations in the two *Cinema* volumes, is "the powerful, non-organic Life which grips the world" (81) such as we encountered, escaping strata and cutting across assemblages, in *A Thousand Plateaus* (507).

Among readers of Deleuze, Claire Colebrook is most explicit in relating the philosopher's cinematic analyses to the question of life, specifically a life of intensity not reducible to the animate. "Deleuze," she writes, "installs technology at the heart of philosophy *and* life.... Only when the human encounters the inhuman will we know what the human body can do, and only when life opens itself up to violence, destruction, death and zero intensity will we be able to discern just what counts as 'a' single life.... It was [in Deleuze's work on cinema] that he showed how creative philosophy and the human brain might become if we confronted our relation to machines."[8] As she explains, to the extent that cinema "thinks" the world as flow, as the life of a plane of immanence in the terms I referred to previously, life is to be understood as perception, which is to say "a virtual power *to relate* and *to image*.... Life is *imaging*, a plane of relations that take the form of perceptions precisely because something '*is*' only its responses. Before there are actual terms—'mind' on the one hand, 'world' on the other—there is a potential for relation, and relations for Deleuze are best described as images."[9] In that way, as she observes elsewhere, cinema provides the potential for "a life of the image beyond the life of spirit."[10] For Deleuze, then, cinema constitutes a technological thinking in images, one that is capable of changing the way the human brain thinks. It does so by unhooking the spectating eye and body from their thrall to an "intellect oriented to fixing the world into so much immobile matter for manipulation and re-organization" and reengaging them as machinic assemblages within "a life that is at one and the same time a capacity for connection, integration, system and meaning... and a tendency towards connections that would open the territory to thresholds that de-form its original system."[11]

The vitalism that Colebrook describes appears in contrast to, or as a radical extension of, another equally legitimate understanding of Deleuze as affirming life in the Nietzschean sense. In his *Gilles Deleuze's Time Machine*, David Rodowick emphasizes how "the time-image lets us believe again in 'life.' This is neither a romantic nor a vague metaphysical concept.... Life is change."[12] Quoting such passages as Deleuze's invocation of the Nietzschean advocacy of "outpouring, ascending life, the kind which knows how to transform itself, to metamorphose itself according to the forces it encounters" (*Cinema 2*, 141), Rodowick makes clear that the stakes of such an affirmation of life involve thinking "what we have not yet become... acknowledg[ing] change and becoming as forces, or the force of time as change."[13] As Rodowick understands it, Deleuze finds that transformative force in a type of becoming-minority of cinema that takes place as the time-image unfolds the linearity of "organic narration" in favor of "a serial organization of images and sounds."[14] Indeed, one of the examples of that tendency that Deleuze emphasizes in *Cinema 2* is precisely Godard's "generalized serialism" (*Cinema 2*, 276), which is based on the principle of an "irrational cut" that "no longer forms part of one or the other image, of one or the other sequences that it separates and divides" (277). The irrational cut (by no means restricted to Godard but common in the New Wave) may involve inserting anomalous images or accentuating the interruptive use of the blank or black screen (249). Its effect is to replace the film's horizontal sequential progression—Rodowick's organic narration from image to image, sequence to sequence—by a vertical or simultaneous serial linkage. Deleuze gives as examples painting in Godard's *Passion* or music in *Prénom Carmen (First Name Carmen)* (276), but as a general rule the disjunction operates most powerfully to imply a "new figure of sound" (249).[15] Sound, as we know, was always in a disjunctive relation to the image track: it came along to be added to the image thirty years into the history of cinema; it has multiple layers of its own (dialogue, ambient sound, music); it often overlaps images and sequences; it has the potential to emancipate itself from the image altogether, as in the films of Marguerite Duras. It might therefore be argued that sound is the paradigm for the disjunctive relationality of cinema, which means, returning to Colebrook's terms, that sound is *the life of the image* or *the image of life*. If relations are best described as images, if life as relationality is all imaging, then the sound that *relationalizes* the cinematic image, or reveals the relationality "within" that image, would in a sense be its life; or, from another perspective, sound would have the capacity to return the image to its relational essence, as it were, return it to the plane of immanence or life.

That will be the argument of what follows, concentrating on the music that begins as, and predominantly functions as, the outside or limit relation to the image. Among elements of the sound track, dialogue and ambient sounds have a logical or diegetic relation to the image track, but nothing in the image calls for a nondiegetic musical accompaniment even if various aesthetic and sociological factors led to that sort of accompaniment during the era of silent film. Music in a talking film is therefore the disjunction in and of the image, as it were, the very disjunctive interstice within it. At least that is what Godard appears to invoke, bringing back onto the plane of the image "the limit or interstice, the irrational cut [that] passes especially between the visual image and the sound image" (*Cinema 2*, 278). From the phenomenological point of view that dominates the theory of film, sound is sensorially registered in the image but could never become visual unless as a synesthetic aberration. Dialogue and ambient sounds, on one hand, would be closer to the visual inasmuch as they issue from characters whose lips we see moving (or whom we identify as belonging to the image) and from objects that similarly belong. Nondiegetic music, on the other hand, is almost always off; it has no visual justification. It is usually made to blend harmoniously with the images, to reflect their mood, but it has no visual place. According to Deleuze's schema, though, all the sound components, including music, are separate or aural "only in the abstraction of their pure hearing" (*Cinema 2*, 235). Within the image, of which they form an integral fourth dimension, "they rival, overlap, cross and cut into each other, they trace a path full of obstacles in visual space" (235). That means that the image is divided or overlaid as much by sound as by visual elements, and its aural components "*do not make themselves heard without also being seen, for themselves, independently of their sources, at the same time as they make the image readable, a little like a musical score*" (235, emphasis added).

Such an apparent reversal of the traditional hierarchy of image over sound, sound's *musicalization* of the image, making it resemble a score, might be better understood in terms of Deleuze's "crystal-image," once again to deemphasize the role of perception or perceptibility in the machinic relations of cinema. For what takes place in the nonchronological time of the crystal-image is a doubling or indiscernibility of actual and virtual that "is definitely not produced in the head or the mind, it is the objective characteristic of certain images" (*Cinema 2*, 69). One such image by Godard might be found in *Week-end*, in an intradiegetic musical performance that suspends visual time and allows the traditional virtuality, or external other face of the image that is music, to reverse sides and, as it were, become actual for the duration of a seven-minute sequence. The 1967 film is one among many—no doubt any—of Godard's first period commercial films to raise the question of the sonic other of the image; it is also notable for being the last film he made before his Dziga-Vertov collective experiment. About two-thirds of the way through the film, the murderous husband and wife protagonists, Roland and Corinne, hitch a ride on a truck driven by an itinerant pianist. In the following sequence, we are treated to an *Action Musicale* (as the intertitle tells us) that is a single (seven minutes eighteen seconds) shot consisting of three, almost 360 degree panoramic sweeps of a farm estate, to the accompaniment of Paul Gégauff playing a Mozart piano sonata. Gégauff's piano has been installed in front of the barn, and his audience consists of a more or less accidental assembly of curious or bored farmworkers, a couple of young lovers, and the protagonist couple. During the first two 360 degree counter-clockwise pans, each of which lasts about two minutes, Gégauff twice plays a one-minute extract of the sonata, stopping at approximately the same point to offer short discourses (fifty-three seconds, thirty seconds) on the place of Mozart in the history of music. Following that, during a clockwise pan, he continues his performance for a little longer than three and a half minutes, up to the end of the sequence.

There is therefore a coherence and unity to the sequence (a "scene" according to Metz's categories[16]), that coherence and unity being precisely provided by the music and the pianist's commentary (the only other dialogue consists of two quick, garbled questions or comments uttered by Corinne). The sonic coherence is reinforced by the fact that for three of its four static moments, the camera pauses on the pianist. The on-screen musical performance would therefore seem to provide the narrative signification of the sequence: *what transpires is the recital.* However, the visual track also offers two different, clearly identifiable cohering structures. The first is provided by a farmworker who is in the frame at the beginning of the take and remains its focus until a point of divergence when the camera leaves him behind, just before it glimpses the couple I have referred to as young lovers walking from background to foreground. The farmworker will reenter the frame almost six minutes later—in the third, clockwise panorama—now carrying a shovel that he seems to have gone to retrieve, ostensibly returning to his work (however much he dawdles), and he will disappear into the hay barn, behind the piano, at the end of the sequence. He therefore reinforces a thematic of

FIGURE 4. Musical action 1 (Anne Wiazemsky). Still from Godard, *Week-end.*

class opposition or conflict that has already been developed in the film, incarnating a proletarian counterpoint to the leisured class (although they are also predominantly rural workers) of idle music devotees. One would have to accept, within the context of the film up to this point, but also in view of the fact that the sequence begins and ends with the farmworker's movements, that his actions constitute a narrative with as valid a claim to importance as the musical performance: *what transpires is that the farmworker goes about his work.*

The contrast between laboring and leisured, or landed, classes is reinforced by the appearance of a second competing visual structure, which works around, and extends from, the couple of young lovers. They appear in the frame—to replace the disappearing workman—in front of the chateau and might be imagined to have emerged from it. They inhabit the foreground for about forty seconds toward the end of the second panorama, approaching the piano to listen to the recital. The female character is played by Anne Wiazemsky, star of *La Chinoise* and Godard's new wife of a few months. She is captured sauntering alone in front of the camera for a few seconds of spectator delectation (Figure 4). If we understand this young couple to provide a contrast with the workman, they also provide a foil for the dominant subjects of this third visual structure, the protagonists Roland and Corinne. The latter are discovered by the camera twenty seconds into its first panorama, he standing yawning (same yawn each of the three times the camera finds him), she sitting, equally bored, so much so that she goes looking for excitement elsewhere, and early in the third panorama, we notice her in the background, sitting at the feet of another man and holding his hand. She rejoins Roland a little over a minute later, entering the frame from the right, and the camera pauses on the two of them for half a minute before again picking up the workman and following him back to the piano and the end of the panorama. Between the younger and older sets of couples, there is reinforced the narrative of bourgeois life: *what transpires is that the bourgeoisie enjoys its weekend.*

The point at which the final pan leaves Roland and Corinne to resume its clockwise sweep coincides with both the reentry into the frame of the workman with his shovel (Figure 5) and the resumption of the piano recital. At the precise moment when we see both the protagonist couple and the workman, and also hear the sonata taken up again, we cannot say which narrative structure is driving the film: our interest in the protagonists, the movement of the workman, or the flow of the music. Music has therefore been invested with a narrative content (and classical music, of course, follows its own classical narrative structures) that to some degree competes with the major diegetic and thematic forces of the film. More importantly, it has come to inhabit the image. For neither can we say with certainty whether it is one or the other narrative structure, or one or the other set of movement-images (all of which inhabit the same camera shot), or something like the time-image of a force of music that is at that point driving the film, indeed constituting it. It is not just that we have seen music being played; rather, we "see" what we hear, to the extent that it is imbricated in, intersects or coincides with—at least at the moment in question—the structures of coherence functioning on the image track. At the 57'18" mark, where we encounter this coincidence or intersection between, on one hand, the workman and the music that we have been following from beginning to end of this sequence and, on the other hand, the couple we have been watching since the beginning of the film, the music can be understood to be *seen* even in the absence of a visual representation of piano and pianist.

Music here—and conversely, I would argue, the image—is therefore

FIGURE 5. Musical action 2 (Roland, Corinne, worker). Still from Godard, *Week-end.*

notable for a type of becoming-modernism different from its fragmentation in many, if not most, other films by Godard. Douglas Morrey has pointed out that the three panoramas with Mozart sonata accompaniment provide a counterpoint to *Week-end*'s famous tracking shot of the traffic jam, road wreck carnage, the former's circularity contrasting with the latter's linearity.[17] Both are of similar length (a little more or less than seven and a half minutes), and both provide a sustained continuity of nonmontage film time (in the case of the *Action Musicale*, it is a single shot, in the case of the tracking shot, an as-if-single shot[18]). Similarly, the relative integrity of the musical interlude contrasts with the cacophonous mayhem of much of *Week-end*'s sound track, in particular the two other on-screen "performances" constituted by the ubiquitous horn blasts of its first half and the drumbeats that are constantly in the background of the Seine et Oise Liberation Front cell scenes. The *Action Musicale* of *Week-end* thus constitutes a specific relationality of image and sound track as well as among image, dialogue, ambient sounds, and music.

However, in the repeated comprehensive panning of the sequence, the spectator also sees the machinery of cinema; one cannot not be struck—experiencing the effects of a certain spatial disorientation—by the fact of seeing the camera turn through 360 degrees. For, if the capacity to record movement provides a certain (tautological) definition of the moving picture, the cinema we are now acquainted with, and expect, derives rather from effects of montage combined with effects of a camera that records movement: that is what, for Deleuze, gives the movement-image. The Lumière Brothers in effect already produced that possibility, without knowing it, by stopping their static camera, acknowledging that if film had the capacity to reproduce the life of the living, it could not or would not preserve that life as flow. (However much they extend that possibility, Andy Warhol's *Sleep* and *Empire,* and ultimately Michael Snow's *Wavelength,* remain within its structural limits.) And as soon as Méliès filmed different scenes with a repositioned camera, film was well on the way to becoming an edited fiction. It would not be long before the basic constituent elements of a camera that stops and starts, dollies and pans, tilts and zooms, or is hoisted on a crane were in play. Godard's recourse, in *Week-end* and in the context of cinema (his cinema in particular but also cinema in general—whether it rely on Eisensteinian montage or continuity editing), to exaggerated tracking and panoramic shots necessarily returns us to a type of naive cinematic apparatus. From that point of view, beyond the visual, sonic, and narrative structures I just outlined, the panoramic musical scene enacts a certain technological constitution of cinema and, I would argue, posits that as a "becoming-music." Within the limits of the sequence itself, such a becoming-music remains actual, and territorialized, at least as long as the music is performed on-screen, within the film's diegetic structures. But its virtuality is, as it were, here put in place, on stage and into play in the indiscernibility of a crystal time-image: one sees the specific relationality of image, sound, and camera movement, and one hears the music not just issuing from a piano being played on-screen but as music being made within that complex and specific cinematic relationality.

There is, however, another performance, what could be called another "sound performance" in *Week-end.* I refer to its second sequence, four minutes into the film, Corinne's recital (the first word of the scene is her interlocutor's command "*recommence,*" suggesting some sort of repeat performance) of the facts of a tryst involving a male lover and his wife. A minute and a half into the nine-minute sequence, we are given the intertitle

"ANAL-YSE," but however much the male interlocutor conducts himself like an analyst, we know him to be the man we were introduced to in the previous sequence as Corinne's current lover, Paul. Besides, if this is an analytic session, it is heavily erotically overdetermined, as she is sitting on the table at which he is seated smoking, in her underwear. The restricted camera movements might allow one to interpret this sequence as even more elemental cinema than the tracking shot of the traffic jam and the repeated panoramas of the Mozart recital, except that we have seen in film after film by Godard reworkings of the traditional shooting of a conversation: most clearly, there is no use of the shot–countershot convention.[19]

Corinne's account of her threesome might be understood to "track" the long flux, either obstructed or unimpeded, of voyeuristic desire, for that is what is staged by the male participant within the narrative, as if in response to Paul's own staging of her account of it—repeatedly perhaps—from the point he enjoins her to "start again" or "rehearse (it for me)." Or we might imagine her recital to produce some sort of virtual pan into the libidinal space of, first Paul, then beyond the 180 degrees of the screen, toward the spectator. Like the Mozart recital, this sequence is also filmed as a single shot, with the camera remaining static, except for limited zooming and panning. Once Paul has asked or told her to begin her narrative, it is Corinne who mostly occupies or dominates the frame, often in close-up, dimly lit and in relative contre-jour, so that the colors of the film are muted almost to monochrome. She is in three-quarters left profile as she recounts her tale, sparing no detail—cunilingus, masturbation, and substantial elements of anal eroticism and a tergo sex, as suggested by the intertitle. After seven and a half minutes, she disappears in search of cigarettes and returns to dominate the frame in much more oblique right profile (Plate 5). For the remainder of the sequence, we see only the back of her head obscuring Paul's face all but completely. What she now retells, still recounting the prompting of the man within the narrative who is directing the scene ("A la cuisine les minettes, à la cuisine"), becomes, progressively more recognizably, a reenactment of the milk and eggs scenes from Bataille's *Story of the Eye.*[20]

The sound track to this sequence also seems to be traditionally conceived, reduced to a dominance of dialogue (or monologue) and accompanying music. But just as the image track frustrates the conventions of the filmed dialogue—being practically devoid of reaction shots, the main speaker either turning her head away from her interlocutor or offering the spectator only the back of her head and shoulders—so dialogue and music function in competition one with the other, such that were it not for the subtitles, one would have only a barely audible suggestion of much of the explicit sexual material. For Corinne's muffled voice is consistently drowned out by the often discordant and strident music when she is explaining details of the more transgressive acts. Once again, there are many other examples of such indistinct or even silent speech in Godard's films, which again leads one to place the sequence in the classic category with respect to his repertoire.[21] If I am insisting on comparing and juxtaposing it with the Mozart recital, it is by reason of its continuous camera and relative calm—it has the tone of an interlude and is immediately followed by the altercation with the neighbors and then the traffic jam—as well as the way in which the music points outside the image. What it "points to" most explicitly in this case is the perverse sexual activity that is being described. It would be impossible to determine whether the scene that is recounted in Mireille Darc's soft melancholic monotone[22] could in fact be represented more erotically in visual terms; that is, after all, the whole challenge, and impasse, of sexually graphic cinema. But in one obvious sense, what occurs is that certain details get represented instead in musical terms. By intervening to replace the dialogue at certain, sexually explicit moments—apart from giving Godard an alibi in a pre–sexual revolution societal and cinematic ambience—the music becomes a type of perceptual sign of the sex that is being unspoken, an "image" of the transgressions that are being portrayed, at least structurally comparable to the classical practice of cutting or panning to a flaming fire instead of showing the consummation of the sex act.

However, Godard is not that prosaically metaphorical, even if one could counter that the discordant music is appropriate to the sex we are hearing about. The music less replaces the dialogue to figure or signify sexual acts than it sounds in a generalized space of excess with respect to the image. Corinne's distracted and detached recounting creates a strange disjunction between the scene's presence as dialogue and its being positioned "off." It is thus already at least doubly absent, off-screen and in the past, as well as detached from Corinne's experience of it. Indeed, in response to Paul's question at the end, "is it true or a nightmare?" she replies that she doesn't know, which adds a further dimension of "irreality." But if Corinne's detachment plays into the complex erotic system of Paul's desire ("Come here and excite me" is his request or command as the sequence ends), of the voice, of voyeurism, and of a generalized erotico-exoticism that, of course, functions also for the spectator, if for those reasons the scene necessarily inscribes structures of distancing and plays on the unattainable object, I maintain that the music ensures that that is not the only way that *the*, or *an*, outside is identified. Its effect, at those points where—again in a way that is comparable to the 57′18″mark near the end of the Mozart sonata panorama—it intersects with, coincides with, or produces an indistinctness of the voice, is precisely that of indistinction, of a signifying imprecision or excess. And at that "level," we can understand a specific form of that intersection or coincidence to be constituted by the name of Bataille. For his text is "heard" only in a muffled or muted way also, by means of the somewhat oblique and unidentified reference to it (unlike, as we shall see, in *In Praise of Love*), as well as through the indistinctness of Corinne's voice and the discordance of the music. Those familiar with Bataille's erotic literature may recognize the episode being quoted in Corinne's narrative, but even if that is not the case, the spectator necessarily has imposed upon them the disturbing effect of a polymorphous perversity, a "contagion" such as Bataille recounts in *The Story of the Eye*, echoing pell-mell through "the void that yawned within us by means of our entertainments with the eggs."[23] In the novel, that contagion or exceeding of the limits is indistinctly a multiplication of examples from the erotic repertory and an overflowing of metonymic and metaphoric signifying processes—culminating in a scene of cruelty that combines much of what is to come in *Week-end*, and more—adding up, especially for those who know Bataille, to an affirmation of sovereign desire, where the subject's triumph is less that of imposing the self than of dissolving into a desiring flux or machine. Of course no music, in the pure or simple sense, however complex it might be, can figure that, and, as I have argued, figuration or signification is no longer what is in play. Rather, music, playing off and across voice and image, at least points in the direction where cinema begins to think otherwise.

Godard's 2001 film *In Praise of Love* explicitly cites Bataille, in at least three forms, recited and reworked through the complex chronology and characterology of the film. In the longest reference, about a third of the way through the film, a quotation from "L'Amour d'un être mortel" is recited and discussed in the context of *Blue of Noon,* during an audition for the vague creative project—formerly a cantata, now perhaps opera, archive, film, sociological study—being developed by Edgar, the film's protagonist. The quotation is something Edgar heard two years ago and the text of which he has now stumbled across at a *bouquiniste's* stall. It deals with the opposition between competing sovereign desires of the State and of love:

> You say love, but nothing could be more opposed to the image of the loved one than that of the State, whose reason [*raison,* also "prerogative"] is contrary to the sovereign value of love. The State in no way possesses, or it has lost, the power to embrace before us the totality of the world, that totality of the universe which is given at one and the same time outside, in the loved one as object, and inside, in the lover as subject.[24]

In an earlier reference (eighteenth minute), Edgar recalls the woman he wants to play in his "film," saying that she offered "a real discourse, about the State, and the impossibility for the State to fall in love." Then, toward the end of the film, which takes place two years previously, Berthe, the woman in question, gives what is in fact the "original" version of the quote, in shortened paraphrased form: "You say the State, but nothing could be further from the image of the loved one than that of the State, whose sovereign reason is opposed to that of love."

In purely analogical terms, given what I have already developed, one could understand the State–love opposition to be that between a certain order or regime of narrative cinema dominated by the image and "music"—not just the opposition between image and sound but between a present, perceived, sensory-motor cinema and a potential or virtual musical "image." In Bataille's terms, the first would be "acquired" as part of a restricted economy, the second "consumed" in an excessive general economy. *Blue of Noon,* as we saw in our previous chapter, stages the tension between those two economies as its first-person narrator appears to decide in favor of erotico-morbid consumption rather than participate, by means of meaningful or coherent political action, in the Catalan uprising in Barcelona. The revolutionary aspiration that the protagonist rejects is incarnated, in the novel, by the character Lazare, understood to be modeled on Simone Weil, who is the subject of Edgar's cantata project and a constant reference in *In Praise of Love*. Edgar's disillusion regarding the cantata, and his pursuit of Berthe, may be interpreted as an increasing sympathy for Bataille's insistence on a complex relation between revolutionary and amorous excess. In referring to Bataille's novel in the audition sequence, for the benefit of the actor who has never heard of the author, Edgar is heard regretting that when people speak of literary representations of the Spanish Civil War, Malraux's *Hope*, a classic novel of engagement, is consistently cited, whereas *Blue of Noon* is ignored.[25]

What we might call the immediate syntax of reference to Bataille in *In Praise of Love*, which would position the three citations as they appear filmically, without considering their more complex diegetic chronology, provides a cinematic staging of the opposition I am describing. For if we compare the three instances, we hear first an imprecise mention (evocation of someone "not very attractive" but who dared to say things), second a contrived recital (the audition), and finally the "original" instance (Berthe's voice). The audition is filmed as is, with an actor given the script and asked to read it; there is no obvious rupture of cinematic continuity. The other two, framing instances, are less transparent.

The first of the two begins with a shot of three characters—a couple and a solitary man (played by Godard)—sitting back to back on two benches on a Paris street at night (Figure 6). The ambient sounds of the traffic and the voices of the couple, whom we see laughing and speaking, are just audible (that ambient sound continues throughout and into the following sequence). The next shot involves a cut to a black screen (between Edgar's voice-over words "State" and "impossibility"), followed by the appearance of the intertitles "de l'amour" (of love) (Figure 7) and "de quelque chose" (of something), part of the expanded version of the film's title, fragments of which have been projected a number of times since the beginning. In the other framing instance, Berthe's voice-over quote from Bataille begins as documentary footage of three men looking at skeletons from the Holocaust stops with a freeze-frame on one of the

FIGURE 6. Framing Bataille, first instance, shot 1 (couple, Godard). Still from Godard, *In Praise of Love.*

FIGURE 7. Framing Bataille, first instance, shot 3 ("de l'amour"). Still from Godard, *In Praise of Love.*

men, who turns and walks toward the camera (Figure 8). That still image cuts to a black screen (between the words "State" and "sovereign reason"), then again to a gramophone turntable with an LP record turning, as the sound track cuts from "Bataille" to Celan reading his "Todesfuge" ("Death Fugue") (Plate 6).[26] During the voice-over, one hears ambient footsteps, presumably diegetic (stiletto heels of the African American assistant). There is no music in any of the three sequences.[27]

FIGURE 8. Framing Bataille, second instance, shot 1 ("Holocaust"). Still from Godard, *In Praise of Love.*

In contrast to those two framing instances, therefore, the middle instance (the audition) is seen in my terms as simple cinema, ordinary cinema. It may precisely be that ordinary cinema, that ordinariness of cinema, whose mise-en-scène Edgar refuses, saying at the end "it won't work," such that we can hear his frustration being expressed not just with respect to the shortcomings of the surprised actor but also with respect to his whole project. The other two citations or evocations of Bataille exceed the economy of ordinary cinema, in the first place by means of the montage of, in each case, three contrasting shots: black-and-white street scene, black screen, intertitle; and black-and-white freeze-frame, black screen, gramophone. In the second place, however, the blank, black screen that occurs as the central shot of both montages exceeds the economy of ordinary cinema by presenting a form of cinema's negation: an anti-image, the image of a lack of image. Now, if we were to impose a contrived structural organization, derived from the three-shot montage, on the three instances of reference to or quotation from Bataille, reading them as three pieces of a discontinuous montage connected by the dialogue track repetitions, the middle instance (the audition) would correspond to the middle shot of the other two instances, namely, the black screen. That is, if there are three instances, and each of the three-shot, outside framing instances has a black image at its center, then the central instance can be read *en abyme* as black, despite its obvious visibility. That would make the ordinary cinema of the audition sequence a nonimage, equivalent to seeing only darkness, there being nothing much to see. Edgar's disappointment reinforces that reading: it's not going to work, the audition isn't working, my film project won't work, cinema doesn't work, one may as well be watching a black screen.

However, another interpretation presents itself: consonant with what I have suggested about the virtuality of music in the image, we might be tempted to read the cut to black with voice-over relating to Bataille as more literally that, a cut "against" the image in the sense of a cut out of the image, into what can no longer be seen; not just into what can be heard instead but rather into or toward an outside of cinema where what is seen is seen as (though) heard. Then it would point to some sort of indomitable sovereignty of the "musical" image analogous to an image of the loved one in opposition to an image of the State, or at least to a relation of cinematic elements whereby virtual and actual images are indiscernibly images of the loved one and of the State (and one cannot help but think that what resonated for Godard in the Bataille quote was precisely the word "image"). It would be important to understand that move toward an outside of cinema, into what I am tempted to call the "music-image," as something very different from an ineffable, musical, or mystical cinema such as is presumed by various commentators who discuss Godard's work, and his use of music, from 1980 on. Laurent Jullier, for example, argues "that *Histoire(s)* is not so much a work of history as one of sensory experience, an attempt to reach the ineffable, mainly through the soundtrack"; James Williams considers that if Godard is silent about music's "increasingly concrete and plastic role in his work," it is because "music *is* a mystery for him. . . . It is ineffable; it simply *is*."[28] On the contrary, I would have that musical outside of cinema understood precisely as history ([*de*] *l'histoire*), in contrast to just a story *(une histoire),* according to the delineation that is consistently made explicit in *In Praise of Love.* That is somewhat how Douglas Morrey would have us understand the "tender and unconventional love story" of the film, in the context of a history that has to be worked at, inasmuch as "the complex montage of Godard's film, which requires the spectator to work to reconstruct a narrative, seeks to demonstrate this sense of history as an active *process.*"[29] If Edgar insists on not representing a story to the extent of having his project fail, it is not because that story existed in a past that cannot now be retrieved; it is rather because he wants to represent history, that story *as* history. And if the difficulty of that story as history derives from its being understood as love, it would be not because love is in opposition to history (some opaque private story versus history as something public and transparent) but precisely because it deals with the very political intransigence of a love that declares itself to be sovereign within an economy of sovereignty that overturns and disables the subject. No simple image, no simple film, no simple cinema can represent that; it won't work. If it can be figured, it is only in a relation between represented and nonrepresentable, which is where the film leads by means of Bataille, into a black screen that does not reduce simply to no image but rather tends toward something like a music-image.

A number of elements reinforce what I am arguing. First, in almost literal terms, the black screen of the first three-shot framing citation of Bataille's text gives way to the words "de l'amour," which inscribe upon the blackness, however temporarily, a signification. (Some) love is to be "seen" here, set against the darkness. We read something of the image of sovereign love, or something of the sovereign image of love, in, through, or on the other side of this darkness. But that signification is quickly qualified by *de quelque chose*: love is not something abstract or ineffable, it comes to be historicized as the particular instance of love of something; the abstraction *de l'amour* comes down to some material level, the level of some thing. Discussions of *In Praise of Love* have not, to my mind, sufficiently emphasized that what is being praised in the title does not remain simply love for longer than forty seconds into the film, and no longer than a momentary shot of blank pages. *Éloge... de l'amour... de l'amour... de quelque chose* is how the title is read, and the words *de l'amour... de quelque chose* are repeated a minute later and at various moments during the film. When the art dealers are speaking of Edgar's project, its title is said to be "something of/about love, 'praise' I think" (quelque chose de l'amour, "éloge" je crois). The words *quelque chose* thus repeat both an imprecision and a qualifying precision that makes the film no more specifically about love tout court, or Edgar's love for Berthe, than it might be, for example, about love for some particular historical thing: this camera's love for Paris, or love for (the history of) cinema; but neither is it more about love through history (Tristan and Isolde, the resistance couple) than it is about love *as* history.

The second reason to read the musical outside of cinema as history emerges via the second framing montage, where the black screen is preceded by a still that evokes the Holocaust (spectators–museum visitors before a heap of skeletons) and followed by a spinning LP record that will come to be heard as Celan reading "Death Fugue." We move from a documentary representation of the Holocaust through black screen to a poetic evocation of it, or at least to the proleptic announcement, on the image track (turning LP), of what will become a poem of the Holocaust. The cut to black is, of course, a cut to a type of contemplation that supersedes and exceeds what a documentary, however well intentioned and historical, can tell us of the Holocaust; recognition of the impossibility of representing it, the impossibility of poetry after it. But, as I read the syntax of the images here, the black screen that appears immediately after we hear the words "th[e image] of the State" goes further than the simple recognition of the impossibility, for example, on the part of cinema, of representing the Holocaust, a failure that we know to be the structuring impulse of *Histoire(s) du cinéma* and, by extension, *In Praise of Love*. In the first place, we are now "looking at" a State whose murderous totalitarian excess perversely rejoins, in Bataille's terms, certain aspects of erotic excess, as is suggested by the end of *Blue of Noon*. We saw in our previous discussion how the Nazi "rising tide of murder" is a proleptic fact of Bataille's novel and how the culmination of the narrator's sexual relations with Dirty—in particular the scene in the cemetery in Trier—forms a stark contrast, and a form of resistance (as well, perhaps, as a morbid attraction), to the increasingly militarized social and political background. In the black screen said to be an image of the State we might be looking at such a history, one that morbidly or obscenely comes back to meet "love" on the other side of it. If there is music in or as *that* history, it would necessarily be a tortured music far from the contemplative lyrical form music adopts in most of Godard's usages of it; music in or as that history would be heard on the other side of such lyricism. A tortured music is, of course, something music can be (as in the strident tones of *Week-end*). Yet, once the image cuts again, we take a further step beyond: Godard suggests by means of the LP, and then by means of Celan, that there remains, miraculously even, miraculously and torturously, the potential for more music, another lyric. The shallow focus of the close-up on the record means that it fades out of focus and into black about halfway up the image, so that the upper half of the screen is still as blankly black as the previous image. We can see this image less as one that comes to be attached by horizontal montage to the previous one than one that is stuck to it or half torn away from it. We see a half-image of previously unseen music—I think it is fair to say that a spinning LP necessarily signifies music—and a half-image of music left over from the black screen. According to this interpretation, the Bataille citation voice-over, conjoined with the cut to black, takes us out the other side of darkness into an image of music, and further still into the music of history recited by Celan.

Last, we can see music as history by means of the very fact of montage, cinema's rhythm of dialectical progression. Godard has used the interrupted time of montage, as exemplified in the syncopated effects we have been examining here, to greater advantage than any other filmmaker. Montage thus becomes, finally, as the film of the same title explains, *our music*. In *Notre musique, notre musique* is *apposed* to the "principle of cinema," which consists in "going toward the light and directing it on our night . . . our music." We hear that spoken by Godard (playing himself) in the course of a lecture he gives in Sarajevo. This is how his voice-over presents it:

> Shot and reverse shot. Imaginary, certainty; real, uncertainty. The principle of cinema: go to the light and direct it on our night . . . our music. (Champ et contrechamp: imaginaire, certitude; réel, incertitude. Le principe du cinéma: aller à la lumière et la diriger sur notre nuit . . . notre musique.)

In the interrupted syntax of that dictum (whose precise punctuation is difficult to determine), music is in fact apposed either to the principle of cinema or to our night. But as is clear from the immediate context of the statement, and in the syntax or narrative of what Godard has been saying in his lecture up to that point, we can understand the principle of cinema to derive from montage, from the more or less dialectical appositions of shot and counter- or reverse shot, by which there is either a simple repetition of sameness that masquerades as difference or as opposition (which is what happens in dominant conventional film—the example shown is from Howard Hawks) or a difference that emerges from an apparent sameness, for example, two news photographs showing opposing sides of a political situation (the examples are Israeli–Palestinian and, as in the disturbing categorization of resistant or compliant concentration camp victims, Jew–Muslim[30]). In one version of the productive application of the principle of montage, there is progress through the dialectic toward a truth about reality (light shed on our night); in another version, one achieves something closer to the sacred, exemplified in the film by the iconic value of the Virgin of Cambray or the martyrdom that the young female protagonist seeks and seems to achieve. Those "sacred" examples are, as it were, beyond vision, at least cinematic vision: the Virgin is said to have "no movement, no depth, no illusion"; the prospect of martyrdom (whose cinematic pedigree comes from Dreyer's *Passion of Joan of Arc*) necessitates closing one's eyes, providing a representation to oneself ("imagining," as the subtitle has it) rather than looking.

Much of Godard's work of the period of these two films, and in particular the *Histoire(s) du cinema* project, involves intense investigation of, and experimentation with, the montage principle. That work already amounts to another cinema, a whole other combination and juxtaposition of images, and of image and sound, an extraordinarily rich and productive reworking of the story and history of cinema. And that reworking figures explicitly as a critique, and an attempt to atone for, cinema's historical failure to represent, or even prevent, Auschwitz. Godard's project would already constitute a transformation of the principle of cinema such as has been enacted in countless mainstream films; it would already constitute an unheard music in and of cinema.

Yet if we attend to the other side of the juxtaposition of our night and our music that Godard produces, precisely our music *as* our night rather than the light cinema seeks to shine on that night, it *sounds* different again. For in the images that accompany the explication of the principle of cinema—"go to the light and direct it on our night . . . our music"—Godard again has recourse to the darkness of a blank screen. At the point at which the words *notre musique* are uttered, the image is blank not just in the sense of dark but rather emptied of the visible. A lightbulb has been swinging back and forth across the screen while Godard emphasizes the effects of the shot–reverse shot dialectic; mostly, it has swung just to the left and right edges of the frame—to the point of the bulb being dissected by those edges—but now it swings out of the field of vision. In fact, though, it doesn't just swing beyond the edge of the frame; it moves more radically outside the range of its pendulum movement, for as it moves, *left to right*, to the edge of the shot, there is a cut to darkness while the words *notre musique* are uttered; then the light reappears *from the left*. On one hand, this looks like a classic Godardian bad splice such as he has been provoking us with since the opening minutes of *Breathless*. But it is more than that: the movement of cinema has brought the image to the limit of the frame of its possibility; its ability to direct light on darkness has been brought to its own limit. As a result, the image has been ceded to darkness. But in that darkness, precisely thanks to it, that is to say in the very rupture that darkness produces within cinema's work of light, from out of the black space between images that is cinema's constitutive possibility, montage emerges. Without it, Godard would insist, the image cannot tell any truth, it cannot shine any light, it cannot measure up to history. But because of it, Godard would seem to suggest, from at least as far back as *Week-end,* as I have tried to argue, the dialectic of juxtaposed images is subjected to its own outside potential, which is an excess beyond any simple opposition. If our music is heard in the film of that name, beyond the ECM extracts that form a rich counterpoint to the film, beyond montage as productive of a photocentric revelation, it is because it is understood also as the montage of our night, something seen beyond vision, something seen heard on the other side of the image.

The montage of our night, of image and music, of image and its outside, would be a montage that "takes on a new sense," a montage of incommensurables such as Deleuze speaks of in reference to Godard (*Cinema 2*, 181–82). Such a montage is no longer simply a juxtaposition but necessarily opens the (incommensurable) interstitial space: that between images, of course, but also within the image, between image and sound, indeed, within and between any of the elements constituting cinema: "The question is no longer that of the association or attraction of images. What counts is on the contrary the interstice between images . . . a spacing which means that each image is plucked from the void and falls back into it. Godard's strength is . . . in making [that] a method

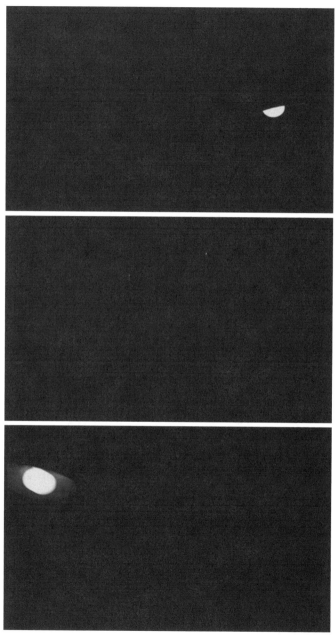

FIGURE 9. "Go to the light and direct it on our night . . . our music."
Stills from Godard, *Notre musique*.

which cinema must ponder at the same time as it uses it" (179). In *Notre musique,* and in "the principle of cinema . . . our night . . . our music," Godard might be said to have achieved a visual montage of music, ceding the image to a type of sovereignty that exceeds it, but by means of its most intimate or immanent technological possibility. "In film," he has said, "there's rhythm, it's more like music, that's how I came to use black for rhythm":[31] the rhythm, therefore, of a varied syntax of shots, whether exaggerated panning and static takes, complicated montage combinations, or the adventures in digital color of these later films, as well as the rhythm, as we have seen, that is punctuated by black. But it is not only in the sense of temporal measure and syncopation that music comes to be incorporated into the visual track. In staging cinema's discontinuity—the inanimation of montage that makes it tick—as a means to have it sensed otherwise, Godard creates for it a different type of flow, making it think again. As a result, there resounds in his films the tonality and beating of nonorganic cinematic life.

9

MEDITATIONS FOR THE BIRDS

Descartes

So, still more or less still life, eighteen years on, stove with melting wax. Descartes is holed up in a room that is, in a sense, all stove or oven *(un poêle)*, composing his *Discourse*. He will publish it somewhat *maskedly*, in French, in Leiden. In it he will relate—also somewhat *maskedly*—what the literature has usually described as his nervousness following Galileo's judgment by the Inquisition, a nervousness that led him four or five years earlier to break off writing his *Treatise on Man*. But when he records the fact of life of that anxiety, to which this chapter will return, should we understand his life, his thinking, and his writing of them to be scrupulously responsible to and for themselves and void of any animal reaction; should we presume there to be no residual automatism repeating and mechanically inserting itself within such an "autobiographilosophical narration" (Derrida, *Animal*, 75)?

I. Imagine an original realistic plush beanbag bird with authentic sounds. Imagine one or more of them in your possession. Fancy, as a result, that "a gentle squeeze is all you need to brighten your day with . . . natural songs." Fancy that. Bask in the joy of it. What will not be so immediately obvious, however, is the fact that you have to repeat the squeeze over and over— many, many times, according to the indeterminable pleasure quotient of each individual—to brighten your whole day. Unless you be one of those cheerful souls whose entire diurnal existence can be brightened by a single avian utterance lasting just a few seconds. Even then, however—and it is here that the problem really begins—it won't be immediately obvious whether what you have heard can in fact be defined as a single utterance.

What you get from that one gentle squeeze sounds like a single repetition of a musical phrase or birdcall. But the sounds produced by one squeeze do not perhaps constitute a single repetition; what one hears may indeed be a single birdcall or a single performance of that call consisting of a single repetition of a series of notes. For example, in the case that will be my paradigm, a certain bright red plush beanbag bird, when squeezed, emits eight "notes" (presuming the standard Western chromatic definition of a note applies), then pauses, then again emits the "same" eight notes. To determine whether what one has heard is one call or two calls, one would have to research the literature and understand differences among call, song, repeat, and serial singing behavior.[1] In the meantime, however, we will doubtless have to accept and interpret the decision of the Cornell Lab of Ornithology, for it is thanks to that institution that one hears the "repetition" just referred to, it being the source of the genuine recorded sound for incorporation in certain toy birds produced in association with the National Audubon Society and Wild Republic.[2]

So, accepting all that, you squeeze once, and the cardinal in question sings once. But this bird's song is half composition and half repetition, or a composition that is pure pleonasm, a series of notes followed by its own redundant and tautological repetition. I have verified that repeatedly, by squeezing not only the cardinal but also the American robin, the common loon, and the blue jay. I consider the number of repetitions of the experiment, that is to say, my repeated squeezing of jay, robin, cardinal, and loon, to be sufficient for me to have scientifically proven the fact of that tautological repetition in the case of the Cornell–Audubon–Wild Republic birds.[3] You squeeze once, but they always sing their song twice they always sing their song twice. Or at least, what they utter once, they utter twice. Or, differently put, they utter once what twice they utter; they utter once when twice they utter.

A whole series of questions comes thus to be raised: is that song a verse or rather a refrain? Are the birds making music or simply mimicking, parroting, or aping themselves? Is their song a call and response, a chant, or simply a repetition? Is it a theme and variation, indeed an improvisation, or rather a mechanical repetition, or indeed reproduction? Those questions might also be tantamount to asking what life is in it: what amount of life in the sense of what form of life comes out of such a squeeze. The answer to that question is complicated, first, by these being toy birds rather than real birds and, furthermore, by their uttering recorded real sounds rather than real live sounds. For we have been led to presume that recorded sound is as opposite to live sound as a toy bird is to a real bird, even though the sound itself is no more perceptually different in one case than in the other. Birds themselves don't have a problem with that, and indeed a live cardinal perched on the television antenna two houses away responded readily both to the squeezed beanbag stuffed toy and to a relatively cheap, flat, and toneless electronic version of a call emitted from a battery-operated baby rocker.

It would be for that reason, of course, that Descartes distinguished, famously in the *Discourse,* between perceiving and thinking, as he also distinguished between producing sounds and speaking a language, and so between an animal, or, indeed, an ingeniously constructed machine, and a human. We might be deceived by a machine looking like a monkey to the extent of having "no means of knowing that they did not possess entirely the same nature," until the moment such a machine tried "to declare [its] thoughts to others." Indeed, even if the machine, and the animal, might be made to, or be able to, utter words, "they cannot show that they are thinking what they are saying" (*Writings I,* 140). A machine might be masked to look like an animal, and an animal might mask its inability to think by uttering words, but we wouldn't be fooled for long. Yet what if the deceptions to which the human is subject at the level of perception did not simply disappear once that human acceded to self-reflective thinking, not only because of the returning repressions of dreaming and madness that have been extensively discussed[4] but also by virtue of returning or recurring itself, of an automaticity in the heart of animal life in general? What if—to put it another way, and to have this final chapter cross-thread back to our first—life itself advanced masked by automatic repetition from the moment of its debut?

For Descartes there is, in the end, no doubt: "there are no men so dull-witted or stupid...that they are incapable of arranging various words together and forming an utterance from them in order to make their thoughts understood; whereas there is no other animal, however perfect and well-endowed it may be, that can do the like.... This shows not merely that the beasts have less reason than men, but that they have no reason at all" (*Writings I,* 140). But Descartes's reduction of the animal to something *a*-linguistic and *a*-rational functions in concert with—and comes across almost as spite for—his recognition of an increasing capacity on the part of not just animals but also machines to imitate life. Indeed, if the Descartes of the *Discourse* seems to want to relegate the animal to the reactivity of an ingeniously constructed machine, it is perhaps because he has, from the time of the *Treatise,* been dealing with his mechanist presumption that the human is a machine, albeit one "made by the hands of God" (139).

Four years later, in the Second Meditation, the problematic is reposed. Descartes is now writing publicly, he is more than twenty years beyond the "*larvatus prodeo*" fragment, and his mask is supposedly off. But the machine masked as human is back. Turning his attention from the wax melting by the fire, the philosopher looks out his window and sees men crossing the square: "I normally say that I see the men themselves, just as I say that I see the wax. Yet do I see any more than hats and coats which could conceal automatons?" (*Writings II,* 21). The French text is here more descriptive than the Latin from which the English translation is drawn: "cependant que vois-je de cette fenêtre sinon des chapeaux et des manteaux, qui peuvent couvrir des spectres ou des hommes feints qui ne se remuent que par ressorts" (*Oeuvres IX,* 25; yet what do I see from this window but hats and coats which could conceal specters or fake men that move only thanks to metal springs). But these human decoys in the form of mechanical androids nonchalantly strolling through seventeenth-century Holland are in good company. For when it comes, a few pages earlier, to his own living body, Descartes imagines something like a moving skeleton with organs and limbs attached, a contraption of bone and flesh that can be best observed when dead: "the first thought to come to mind was that I had a face, hands, arms and the whole mechanical structure of limbs which can be seen in a corpse" (toute cette machine composée d'os, et de chair, telle qu'elle paraît en un cadavre; *Writings II,* 17; cf. *Oeuvres IX,* 20).

If Descartes is given to imagining or finding a lifeless machine beneath the human form, it is again to prove that he possesses powers of thinking that transcend those of the imagination or "what they call the 'common sense,'" such as are available to any animal (*Writings II,* 22). Any old animal can believe what it sees; only a thinker can deduce what it sees to be other than what it is, for example, either dead or alive. Descartes would have us believe that no faculty of nonhuman animal sense perception can lead a lowly creature to determine whether coats and hats, seen from a distance, hang on live humans or inanimate automatons. No such faculty will, conversely, permit the lowly animal to know a piece of wax as anything other than a piece of wax. Or, in the precise terms of a Descartes not at all confused by the question of what is really alive enough to put on a hat and coat, only a thinker can undress a piece of wax: "Any doubt on this issue would clearly be foolish; for what distinctness was there in my earlier perception? Was there anything in it which an animal could not possess? But when I distinguish the wax from its outward forms—take the clothes off, as it were, and consider it naked [*mais quand je distingue la cire d'avec les formes extérieures, et que tout de même que si je lui avais ôté ses vêtements, je la considère toute nue*]—then although my judgment may still contain errors, at least my perception now requires a human mind" (*Writings II*, 22; cf. *Oeuvres IX*, 25).[5]

According to certain observations and deductions in Descartes's Second Meditation, therefore, the capacity of the human mind, in going beyond sense perception and the common sense, in contradistinction to the capacities of the animal, progresses as follows: it sees its body stripped down to a lifeless machine; it understands that in order not to remain at the level of seeing, which might perceive real clothes hanging on lifeless automatons, it needs to employ cogitational judgment; and finally, such judgment or thinking raises the human above the animal by undressing a piece of wax. At best, there would seem to be some hesitation, not to say confusion, over how to describe the life-form that is reason; at worst, in order to come to be by thinking, one would be required to "suspend or rather detach, precisely as detachable, all reference to life, to the life of the body, and to animal life" (*Animal*, 72). One *is thinking* only against that background of *inanimation*.

II. Descartes's meditational mantra wax, from the category of "bodies which we touch and see," begins within the artisanal ambit of the animal, having "not yet quite lost the taste of honey." It is still very much bee's wax. Once it is put by the fire, it is found to be "not after all the sweetness of honey, or the fragrance of the flowers, or the whiteness, or the shape, or the sound, but rather a body which presented itself to me in these various forms a little while ago, but which now exhibits different ones. . . . Let us concentrate, take away everything which does not belong to the wax, and see what is left: merely something extended, flexible and changeable" (*Writings II,* 20). What the wax loses in order to become something conceptualized by the mind is, among other things, its animality, in the sense of the knowledge of it that derives from smell and taste (honey) and hearing (the sound it made, while still solid, when rapped with a knuckle). Before being undressed, or in the process of being undressed, the wax will be divested "of its sensible finery [*parures*] or facing [*parements*]; namely of what, in it, remains animal or exposed to animality" (*Animal,* 73).

Even though Descartes is trying to argue for knowledge of properties in wax, or properties of the object or body in general that do not derive from perception, when he comes back to the possibility that "knowledge of the wax comes from *what the eye sees,* and not from the scrutiny of the mind alone" (*Writings II,* 21, emphasis added), he seems to have acknowledged that the concepts of extension, flexibility, and changeability are owed to a primary visuality. Paradoxically, it is to counter that, as if to distract from the persistence of *vision,* that he decides to shift his attention from looking at the wax to glancing out the window. And it is the *image* of automatons with which he is subsequently confronted that will lead him to renounce vision in favor of judgment; it is as if to replace eye-work by mind-work, one must succumb to the haunting of automation. If elimination of what might be called the animal (and even vegetal) sensory field—honey scented with flowers—is not enough to prove a "perception" derived from "purely mental scrutiny," and if a visual form of life still reasserts itself as the means of knowledge beyond the other senses (the conclusion "that knowledge of the wax comes from what the eye sees"), then such an illusion will be disabused by raising the specter of automatic men. First remove the animal; then, if that doesn't work, introduce the androids. That is how one can isolate a truly human thinking being.

What are we to make of extension, flexibility, and changeability, in fact? On one hand, they are the basis for this distinction between what is known by the "imagination," or sense perception, and what is perceived by the mind and so defines the proper conceptions of a body, "everything...located outside me" (*Writings II,* 22). The extension, flexibility, and changeability of the object as scrutinized by the mind should be understood to transcend all of the following: what is known when one first sees a piece of wax, then sees it melting and spreading out; what is known when one first sees the piece of wax at hand, inside, and then looks out the window to the hollow men; what is known when one first sees wax, solid or melted, then looks out to hats and coats, concealing real men, or springs and sprockets; and what is known when one first *imagines* on the basis of the senses, and then *thinks* on the basis of the mind. But in each of those cases of knowledge, and in the relations among them, extension, flexibility, and changeability would also seem to be involved: visual perception extends from fireside to strolling burghers; more or less clockwork bodies flex, albeit with less fluidity than wax; sense perception changes into thinking. If extension, flexibility, and changeability are not properly speaking properties of the object, not what the object reduces to once its animal sensibility has been removed and once it has been threatened with the absolute lifelessness of automatism, but are rather the conceptual properties by which the object comes to be defined by the mind rather than by the senses, then nothing prevents the mind from extending, flexing, and changing in its turn. However, if we are to avoid watching the mind devolve back into an object or body, we would presumably have to find a way to describe those mental transformations that did not depend in any way on visuality as representative of the senses in general; we would have to understand an extension whose concept eluded or excluded visuality.

III. Though it would be absurd to downplay the role of visuality in non-human animal sense perception, clearly scent and sound function there at a far higher level than they do in the human sensorial hierarchy. A wooden decoy can stand in for a duck, and no doubt the original realistic plush beanbag toy would be able to function as a visual simulacrum of the cardinal, robin, loon, or jay. But we imagine that a bird conceives of a body differently from human understanding beyond those simple visual parameters and that that difference derives at least in part from the fact of the animal assigning a different perceptual and cognitive function to sound. We might also imagine that it is by singing, among other means, that birds self-extend. In the first place, what does that do to their status, in Descartes's terms, within the category of everything located outside me? Does the bird change and extend by means of song in the same way that a piece of wax does? Even he, I suspect, I think, would find the analogy to be somewhat perverse, although we know that he has, in the *Discourse,* allowed for automatic animals—"machines [having] the organs and outward shape of a monkey or some other animal that lacks reason" (*Writings I,* 139)—animals we should presume to be as inanimate as a piece of wax, just as he here allows for spring-loaded automatons in hats and coats crossing the village square. In the second place, what does a bird's sonic self-extension do to its status as a (non-)thinker if it is thereby able to know and understand something like territory—surely a question of the extension of bodies—in a way that we cannot conceive of doing?

Birds use song to mark territory. But we would have to understand the birdsong, even in its territorial function, as different both from pre- or protolinguistic utterance and from communication as we normally conceive of it. To the extent that it is a matter of proclaiming the bird's radius of influence to whatever other animals are to be found within earshot, it has no known or presumed, no specific, addressee. In Derrida's terms, it is a generalized dissemination,[6] in Deleuze and Guattari's terms, a deterritorialization: "Sound owes [its] power not to signifying or 'communicational' values (which on the contrary suppose that power), nor to physical properties (which would privilege light over sound), but to a phylogenetic line, a machinic phylum that operates in sound and makes it a cutting edge of deterritorialization" (*Plateaus,* 348).

Sound's cutting edge is, however, blunted by the *ritornello* or refrain. The refrain is the "tra la la" that characterizes the childhood lullaby or round, Proust's Swann's Vinteuil's little phrase, and the birdsong alike. As such it forms a basis for music in general: "We would say that the refrain is . . . the block of content proper to music. A child comforts itself in the dark. . . . A woman sings to herself. . . . A bird launches into its refrain. All of music is pervaded by bird songs" (*Plateaus,* 299–300). But the role played by the refrain is ambiguous, for though it is a necessary structuring force thanks to which sound becomes music, it is at the same time the limitation from which music must emancipate itself: "We are not at all saying that the refrain is the origin of music, or that music begins with it. . . . The refrain is rather a means of preventing music, warding it off, forgoing it. But music exists because the refrain exists also, because music takes up the refrain . . . in order to take it somewhere else" (300). On one hand, therefore, the ritornello works to territorialize and, when reduced to its simplest comforting or uniting function, exposes music's black hole: "Since its force of deterritorialization is the strongest, [sound] also effects the most massive of reterritorializations, the most numbing, the most redundant.... Colors do not move a people. Flags can do nothing without trumpets.... [Hence the] potential fascism of music" (348). On the other hand, music is able to reinvest the refrain with its deterritorializing capacity and have it reconnect with the unthinkable cosmic forces that are its outside possibility: "Music submits the refrain to this very special treatment of the diagonal or transversal, it uproots the refrain from its territoriality. Music is a creative, active operation that consists in deterritorializing the refrain" (300).

The birdsong alternates in a similar way between music and refrain. It represents a powerful gesture of territorialization—"in animals as in human beings, there are rules of critical distance for competition: my stretch of the sidewalk" (321)—a molar tendency in contrast to which one can posit the molecular noise produced by insects: "the bird kingdom seems to have been replaced by the age of insects, with [their] much more molecular vibrations, chirring, rustling, buzzing, clicking, scratching, and scraping. Birds are vocal, but insects are instrumental: drums and violins, guitars and cymbals" (*Plateaus,* 308). However, resorting not to anthropomorphism but what they call rather "geomorphism" (319), Deleuze and Guattari recognize in birdsong a potential for musicianship that once again rejects the reductive force of the refrain in favor of more creative departures: "Are there not, as Messiaen believes, musician birds and nonmusician birds? Is the bird's refrain necessarily territorial, or is it not already used for very subtle deterritorializations, for selective lines of flight? ... Does it remain territorial and territorializing, or is it carried away in a moving block that draws a transversal across all coordinates—and all of the intermediaries between the two?" (301–2).

So what is there in the repetition of a refrain that causes it to lean one way rather than the other, toward deterritorialization rather than reterritorialization? For Deleuze and Guattari, the answer lies in the capacity to hazard "an improvisation" (311), a capacity recognized in birds, in particular as their simple territorial sweep or sway gives way to the transversal effects of courtship, sexuality, sociality, and beyond. No longer do they simply repeat their calls like a beacon foghorn, warning off competitors or enemies; instead, they become musicians, introducing style: "What objectively distinguishes a musician bird from a nonmusician bird is precisely this aptitude for motifs and counterpoints that, if they are variable, or even when they are constant, make of them something other than a poster, make of them a style, since they articulate rhythm and harmonize melody" (318). Once that aptitude comes into play, the birdsong escapes from the ambit of ethological behavior, and indeed from the ordinary ambit of life, a life that is itself—like the birdsong— both "a particularly complex system of stratification and an aggregate of consistency that disrupts orders, forms, and substances" (336); it rejoins instead the machinic phylum that, as we saw in the introduction, is pre- cisely life beyond organic strata, life understood rather as technological and metallurgical flow. And that sense of metallurgical life is now related to music, for whereas "matter and form have never seemed more rigid than in metallurgy," there is a tendency "within both arts to bring into its own, beyond separate forms, a continuous development of form, and beyond variable matters, a continuous variation of matter: a widened chromaticism sustains both music and metallurgy" (411).

In that way, the birdsong extends, in its most improvisational mode, into musical metal or metallic music. It is an extension that operates a transversal line of flight: no longer about the simple extension of territory— still perhaps having something to do with the flexibility of territory, indeed its changeability—no longer relating by any means to how those proper- ties manifest themselves in Descartes's heated wax. It extends into a very different configuration of both the cultural and semantic dimensions of sound: into differences between human music and animal sound, between music and sound within the nonhuman animal kingdom, and between music and metal in the cosmos of nonorganic life.

For ornithologists, the white-crowned sparrow, like the ovenbird, the Chingolo sparrow, the European redwing, and the splendid sunbird, has a single song. At the other end of the scale, the nightingale is reputed to be capable of up to two hundred different songs, and the brown thrasher, more than two thousand. It is even possible that a male sedge warbler never repeats exactly the same sequence of sounds twice during his lifetime. Some birds with a small number of songs seem capable of doubling their repertoire by singing either an accented or unaccented song type, one for territorial defense and the other for mate attraction. Indeed, etholo-gists believe that the latter two purposes constitute the grand taxonomic distinction for all songs.

There is far from universal agreement, however, concerning what constitutes a different song and hence concerning the repertoire of each species. To take a single example from among my original realistic plush beanbag examples—jay, robin, cardinal, and loon—the cardinal is said by one authority to have a repertoire of eight to twelve songs[7] and by another to produce an innumerable quantity of songs.[8] For purposes of squeezing, the Cornell Lab of Ornithology reduces that variety to a single "whoit, whoit, whoit, whoit, whoit, whoit, whoit, whoit" repeated twice (which may also be a single "whoit" repeated sixteen times). What the cardinal is in fact singing sixteen times, according to the accepted version of the catchphrase as verbalized, Anglicized, and anthropomor-phized in ornithological literature, is "what cheer." That differs from the robin's "kill 'im, cure 'im, give 'im physic" and the blue jay's "Thief! Thief! Thief! Thief! Thief! Thief!" which is said to be "too strident to appear beautiful to the human ear."[9] It is hard to know whether any of my spe-cies could be called "not only virtuosos but artists" in Messiaen's terms, cited by Deleuze and Guattari (*Plateaus*, 316–17). For that talent seems in the final analysis to imply less a quality of improvisation or composition than a quality of performance, the ability to sing better than a competitor. And it seems also to be identifiable in the first place in territorial songs, hence in the fixed, placarded, identitarian, potentially fascist repetition of the ritornello. Hence the quality of a performance might ultimately be determined by its quantitative superiority, by the number of simple repetitions, by a form of mechanical reproducibility; he sings loudest who sings most, longest, or last.

But, to return to Descartes's terms of reference, there might indeed be some analogy between the visible extension of wax and sonic extension such as birdsong, indeed, a difference between melting wax and undressed wax. Melting wax could be compared to competition over territory: the further one is able to extend oneself, however thinly one is finally spread, the greater one's territorial reach, even to the point of the liquefaction into virtuoso song or evaporation into artistic composition. But once it comes to courtship, a bird no longer simply extends; it undresses oneself, attains the property, principle, or concept of extension all the way to communication with another. And in dancing and mating beyond even that, beyond extension, beyond thought, indeed in singing beyond any behavioral program, the bird deterritorializes into cosmic music such as Descartes could never possibly even dream.

IV. Would we prove the existence of God if we were to conceive of him singing like such a superextended cosmic bird, but fallen silent?

After all, Descartes's ontological argument has its problems. In the Third Meditation, he famously insists that his ideas have to have originated elsewhere. He then goes on to categorize those ideas in terms of his own self-representation, followed by "God, corporeal and inanimate things, angels, animals and finally other men." That list is reduced to two categories where the first (the ideas of men, animals, and angels) is found to be derived from the second (myself, corporeal things, and God). However, he then excludes things from the latter group because "I can see nothing in them which is so excellent as to make it seem impossible that it originated in myself" (29). That leaves, therefore, himself and God as the source of all ideas. Continuing in the same vein, he surmises that almost everything could derive from him, everything, that is, except the idea(s) of God: "a substance that is infinite, eternal, immutable, independent, supremely intelligent, supremely powerful, and which created both myself and everything else (if anything else there be) that exists. All these attributes are such that, the more carefully I concentrate on them, the less possible it seems that they could have originated from me alone. So from what has been said it must be concluded that God necessarily exists" (30).

Within this logic, birds (animals) derive from wax (things) which derives from me who derives from God. God self-extends infinitely yet remains immutable; however extendable he be, he is neither changeable nor versatile and therefore not wax; and, since ideas of animals are still more derived than those of things, he is definitely not a bird. When yesterday's Second Meditation wax returns in the Third (we understand it to be the same wax today as it was yesterday, and we wonder whether, if honey is the sole food that never spoils, that gives a certain immutability to wax?), the reader is given a modified version of yesterday's concept of it, which was determined by extension, flexibility, and changeability; Descartes now adds substance, duration, and number. But, he opines, "as for all the rest, including light and colours, sounds, smells, tastes, heat and cold and the other tactile qualities, I think of those only in a very confused and obscure way, to the extent that I do not even know whether they are true or false, that is, whether the ideas I have of them are ideas of real things or non-things." Among those varied qualities, heat and cold will become the exemplar of "material falsity," the representation of nonthings as things, because "the ideas which I have of heat and cold contain so little clarity and distinctness that they do not enable me to tell whether cold is merely the absence of heat or vice versa, or whether both of them are real qualities, or neither is" (30). Where does that leave "all the rest, including light and colours, sounds, smells, tastes?" Sound, for example, and in particular: is it so unclear and indistinct that we are to understand it as merely the absence of silence? Music as the absence of noise? And where does that leave the birdsong? As the absence of rational speech, we might expect a Descartes to respond, but then language and song would be like heat and cold, and we would have no more certainty of the clarity and distinctness, or even of the reality of one (language), than we would of the other (the birdcall). Perhaps, instead, we should understand a birdsong as the absence of what some observers have identified, for example, in the blue jay, as "primitive" or "vestigial": "The bird has the head uplifted in a song pose, and produces a series of mixed warbles and twitters which carry only a short distance and are altogether different from the ordinary noisy calling of the bird. . . . I regard this singing effort of the Blue Jay as primitive, that is, as an indication that the Blue Jay's ancestors were real singers. In short, the bird at certain times reverts to the ancestral song. But such a song . . . has no significance for the present life of the bird."[10] The hypothesis is of an ancestral blue jay with a different repertoire, a blue jay that once lived a whole other prior life, where it related wholly differently to territory and mating; or else, of a real singer cosmic Ur-bird, wholly song, pure, perfect, perhaps infinite musical expressivity, the god of all birds to which today's humble blue jay can compare itself only in the mode of imperfection, but proving thereby that such a total song, ethereal metal, and God as that song, exists even if he can no longer be heard.

V. There is a type of reciprocity in Descartes's movement though angels, animals, and men; through things; to me; and finally to a God who is everything. We should read it that way if we are not to advocate—in the robes of the evil demon's advocate—that there is a confusion of deductive and inductive logic or a circularity of argument from the generality of life and things to the particularity of God as perfect and infinite generality. We should instead hear angels, animals, and men breathing or singing the idea of them through things to me and on to God, who breathes back into or sings back into me, us, them, everything—a call-and-response bird chorus of ideas flooding the universe with song.

Response, after all, is everything. As we noted beginning in chapter 1, as long as response can be differentiated from reaction, it is everything, and, as much as thinking, it is what we are, what makes us human. That will have been presumed from Descartes all the way to Lacan, as Derrida emphasizes, returning to the terms of more than one of our previous discussions: "what never even crosses the mind of any of the thinkers . . . on the subject of response, from Descartes to Lacan, is the question of how an iterability that is essential to every response, and to the ideality of every response, can and cannot fail to introduce nonresponse, automatic reaction, mechanical reaction into the most alive, most 'authentic,' and most responsible response" (*Animal,* 112). Descartes plays the response card at the end of section 5 of the *Discourse,* where we are asked to imagine an automaton with "the organs and outward shape of a monkey or of some other animal," one that "bore a resemblance to our bodies and imitated our actions as closely as possible for all practical purposes," and we are invited to fancy its being "so constructed that it utters words" and even "cries out that you are hurting it." However, Descartes assures us, "it is not conceivable that such a machine should . . . give an appropriately meaningful answer [*répondre*] to whatever is said in its presence, as the dullest of men can do" (*Writings 1,* 139–40). Derrida finds another explicit formulation of the dilemma, and of Descartes's response to it, in the philosopher's March 1638 letter to an unknown addressee. That explicitation, however, involves a rather more elaborate rhetorico-fictive frame, for in the letter, we are presumed to understand the nonresponse of a well-disguised and even impassioned automatic animal based on the experience of a man who spent his life empirically ignorant of all animals other than humans but nevertheless found a way to construct the automatic animals in question. The scene is too fabulous not to quote in extenso:

It seems to me that one must consider the judgment that would be made by a man who had been nourished all his life in a place where he would never have seen any animals other than men, and where, having given himself wholeheartedly over to the study of mechanics, would have manufactured or helped to manufacture several automatons, some of which would have the figure of a man, others of a horse, others of a dog, others of a bird, etc., and which walked and breathed, in short which imitated as much as possible all the actions of the animals they resembled, without omitting even the signs that we use to make known our passions, such as crying out when they are struck or fleeing when someone makes a loud noise near them. (*Oeuvres II*, 39–40, my translation)

And Descartes's response to it repeats, as if reactively, the two means explained in the *Discourse*: "never, unless it be by chance, do these automatons respond, either with words or even with signs, concerning what is asked of them," and although their movements may be more certain and regular than those of wiser men, "they are nevertheless lacking in certain things that they would have to perform in order to imitate us."[11] Derrida comments,

> The scene and logic of the argument seem to me more strange than has been most often noted. Here we have a character, a man, and this man is a man who, having learned, fictitiously, to manufacture impeccable automatons, would conclude *in reality*, by means of a judgment, that the animals are *in truth*, for their part, automatons, automatons of flesh and blood. And why is this so? Because they *resemble* automatons that *resemble* humans. (*Animal*, 83)

So there is no room for debate: however uncannily automatons might be able to imitate animals, or react to pain or noise like animals, like animals never would they be able to respond. Never would an animal, never would an automaton, and especially an automatic animal, be capable of replying to a question. Ask a lifelike automatic animal—created by a fictitious man who has never seen an animal other than a man—whether it is real, and its bafflement would be owed not to the bizarre context of the question but rather to its inability to respond. It would not be that there was no good answer to the question; rather, there was no answer at all.

In contrast to the situation of the *Meditations,* though, and to the idea of clockwork hats and coats crossing the square, in the context of the letter we are at least dealing with sound and utterance, not just visual experience. Yet when Lacan raises the question of animal language and response as one of gesture, he does so firmly within a regime of visuality. In "The Function and Field of Speech and Language in Psychoanalysis," it is a matter of the dance of the bees and the code by which they indicate the direction of and distance to the nectar. But their dance doesn't constitute a language because of the "fixed correlation of its signs to the reality that they signify."[12] The dancing bee may well accurately indicate to its fellows exactly where honey is to be found, but "its message remains fixed in its function as a relay of the action, from which no subject detaches it as a symbol of communication itself."[13] For Lacan, therefore, bees dance *in code,* to which other bees *react,* in contrast to humans, who speak *in language,* to which other humans *respond.* In "Propos sur la causalité psychique," we learn further how the maturation of the gonad in the hen-pigeon is a reaction to the sight of a fellow creature of either sex, even as a mirror reflection.[14] Beyond his surprise at "the purity, the rigor, and the indivisibility of the frontier" that Lacan appears to install between reaction and response, Derrida asks more specifically how that can be so when, "especially when—and this is singularly the case for Lacan—the logic of the unconscious is founded on a logic of repetition which, in my opinion, will always inscribe a destiny of iterability, hence some automaticity of the reaction in every response, however originary, free, critical [*décisoire*] and a-reactional it might seem" (*Animal,* 125).

The destiny of iterability functions, in Descartes and in Lacan, as a technological drift or countercurrent. Descartes's uncanny compulsion is by now familiar to us: mechanical or automotive motifs keep recurring in the *Discourse, Meditations,* and *Letters* like a return of the repressed. It is as if wherever there is an animal—or often even a man—the chimera or fiction of an automatic simulacrum is never far away. The connection seems to function in him like some automatic reaction, the haunting, precisely, of an inevitable automaticity. In Lacan, as Derrida reads him, a relay of visual stimuli and behavioral reaction is permitted to relegate the animal to a life without the unconscious, while neglecting the extent to which the very concept of the unconscious relies on an economy of automatic reactive repetition such as was identified in the animal. But if the destiny of iterability installs a logic of repetition or automaticity in the unconscious, it also extends into the mediation by a simulacrum that is the mirror stage. There is something strangely analogous, even if the analogy be a reverse one, between the coming-to-identity of the human as lack and misrecognition by means of the specular image and the coming-to-cogito by means of the misperception through a Dutch window of well-dressed automatons.

So, pace Cornell, Descartes will have foreseen the original realistic plush beanbag birds with authentic sounds. Even a well-trained artisan who has never seen an animal other than other men, provided he has given himself wholeheartedly over to the study of mechanics, could have dreamed them up. I don't know whether the songs my birds sing bear witness to their passions, and their utterances don't appear to be cries caused by their having been struck, even if I have already suggested that "gentle" is probably not a scientifically rigorous word for the firm squeeze required to make them sing. All the same, what they utter or emit are indubitably automated sounds, manufactured to the extent of being recorded; neither live nor *en direct,* these birdsongs remain irrevocably technologized in one side and out the other of their authenticity. They are the songs of live birds become dead, mechanically reproduced sounds. A living bird hearing their sound may well respond to them, but no one who stopped to speak or to sing to these birds would opine for very long that they were alive.

But what if something in the original realistic plush beanbag birds were to revive them; what if that something were precisely the repetition of their song? What if, even before the Cornell ornithologists decided to repeat the sequence of bird sounds—according, one hopes, to some better-founded scientificity than that inspiring Wild Republic's marketing department's choice of words—what if, even before the song were transmitted and translated into a code of territorial protection, or a code of mate attraction, there were in it sufficient autopoiesis, autoaffection, and autokinesis to constitute nevertheless a form of life? For, as Derrida insists, it is no longer so easy to distinguish that life from what we presume to be its inanimate opposite, to distinguish automotion from automatism. His objection to Lacan is not that there are no parameters or criteria for distinguishing between reaction and response, between code and language, and between animal and human: "Far from erasing the difference—a non-oppositional and infinitely differentiated, qualitative and intensive difference between reaction and response—it is a matter, on the contrary, of taking that difference into account within the whole differentiated field of experience and of a world of life forms" (*Animal*, 126).

How much life is there in a birdsong technologically separated from its living voice? How does that change if it sounds sufficiently alive to have another animal respond to it? *And especially if, as a result of sounding sufficiently alive to have another animal respond, the recorded song stimulates gonadal maturation that will eventually lead to a mating process and the reproduction of life?* Do we make inalienable distinctions between reaction and response and between animal and human precisely not to have to deal with those questions? So that we can reassure ourselves that we—we who know what a response is and how it is different from a reaction—would never be so naive as to respond, in the sense of starting up a conversation with a recorded voice? Even if, when that recorded voice is a voice that sings, on vinyl, on the airwaves, or on the iTunes ethers, we are all too ready to sing back to it or along with it, ever so automatically?

The mechanical monkeys of the *Discourse* appear in that 1637 text, as I mentioned earlier, in the context of Descartes's earlier *Treatise on Man*, a summary of which makes up much of the final two sections of his essay. The *Treatise* has come down to us in truncated form only, but it details the workings of a human body that Descartes supposes "to be nothing but a statue or machine . . . made by the hands of God" (*Writings I*, 99). Thus, behind every fictional manufacturer, having given himself over wholeheartedly to the study of mechanics, whom we encounter in Descartes, there stands the transcendent divine artisan; and every separation of body from mind is reinforced by a determinate opposition between animate life and inanimate machine. But elaboration of that opposition inevitably calls the distinction into question, and the machine of the body often seems to invade the mind. For example, *The Treatise on Man* compares the nerves of the brain machine with pipes in the works of fountains in the royal gardens, "its muscles and tendons with the various devices and springs which serve to set them in motion, its animal spirits with the water which drives them" (100). In that sense, Descartes is already performing the task Derrida sets for us of "taking this grand mechanicist—and what is also called materialist—tradition back to the drawing board" to the extent of reinterpreting "not the living creature called animal only, but also another concept of the machine, of the semiotic machine if it can be called that, of artificial intelligence, of cybernetics and zoo- and bio-engineering, of the *genic* in general, etc." (*Animal*, 76). By the same token, he is led to problematize the generic purity of the modern philosophical and scientific discourse that he has inaugurated. Not only are man and animal persistently shadowed by the machine but Descartes's frequent recourse to fictions such as, precisely, the robotician without experience of animals, and his propensity for fables in general, function as an automatic mechanist penchant within that discourse, as it were his own little artisanal fiction industry manufacturing fabulous machines that both *animate* and *respond to* his reasoning.

Furthermore, response itself comes to be something of a generic industry within Descartes, presuming we are able to distinguish it from reaction. Was it reaction or response, we may now ask again, that led him not to publish his treatise? For as he writes in the *Discourse*, he made that decision after learning that "persons to whom I defer and who have hardly less authority over my actions than my own reason has over my thoughts" (*Writings I*, 141) disapproved of the work of someone else. We know that someone else to be Galileo and we know the persons to whom he euphemistically defers to be the Most Eminent Cardinals of the Commissionary General of the Inquisition, led by Cardinal Maculani. They would therefore be persons likely to *react* to his *act*-ions as quickly as his reason responds to his thoughts. When he wrote that, he could well have been remembering how he once held his own hand to the fire and thereby learned something of the similarity between the human body and a body of wax. Looking forward to the account of it he is to give in the *Meditations,* react he well might. And reflecting on it now, next to his stove, watching the wax melt, he might well call to mind not only the judgment against Galileo but the capital sentence imposed on Giordano Bruno at the outset of the century. Knowing what fire can do to the human body, he would have to presume that, given sufficient heat, flesh and fat would revert to a chemical animality that would have them react, in extenso, as automatically as the undressed bee product of the *Meditations.*

So respond, or rather react, he well might. The Galileo question is the prime concern of three letters to his faithful (like a dog?) interlocutor (respondent? reactant?) Friar Mersenne written between November 1633 and April 1634. Each time he makes his position clear: "But for all the world I did not want to publish a discourse in which a single word could be found that the Church would have disapproved of," or again, "I have decided wholly to suppress the treatise I have written and to forfeit almost all my work of the last four years in order to give my obedience to the Church. . . . I seek only repose and peace of mind," or again, "Though I thought [my arguments] were based on very certain and evident proofs, I would not wish, for anything in the world, to maintain them against the authority of the Church. . . . I am not so fond of my own opinions. . . . I desire to live in peace and to continue the life I have begun under the motto 'to live well you must live unseen.'" He writes the motto, which he repeats after Ovid, in Latin, presumably for the benefit of the cardinals: "bene vixit, bene qui latuit."[15] One wonders how well a parrot would be able to ape those words in the face of the threat of the fire, or how much one would have to slur "bene vixit, bene qui latuit" for the words to become the "whoit, whoit, whoit, whoit, whoit" of another cardinal we have already encountered. One that you sometimes have to squeeze too hard for comfort, though, granted, not as hard as thumbscrews. Be that as it may. One would nevertheless be hard-pressed to make Ovid's quotation resemble the "what cheer" that the Anglophone anthropomorphic cardinal is supposed to utter when it sings.

Ovid's motto can in fact sound or sing as consonantly in English as in Latin: he lives well who lives latently. Try to stay under the radar. Keep the mask on. Descartes is heard uttering paraphrastic versions of the same adage or refrain in the *Discourse* and the *Meditations,* when he praises the Dutch peace (*Writings I*, 126), when he defers to the Church (*Writings II*, 6), when he insists on how small an intellectual territory he seeks to control. My point is by no means to question his courage or resolve, for part VI of the *Discourse* is not without audacity in its irony toward "those whom God has set up as sovereigns over his people or those on whom he has bestowed sufficient grace and zeal to be prophets" (*Writings I*, 142). Nor do I wish to portray him as reduced to parroting a cardinal; rather, I mean to suggest that, whether it be a matter of his words and actions upon hearing of Galileo's misfortunes or some other preemptive maneuver on his part, the more reflexive—or *genuflective*—his response, the more we might tend to interpret it as reaction.

However, "response [*réponse, responsio*]" is explicitly the word he wants to describe the machinery of debate that he establishes after completing the *Meditations*. He invites "objections" to which he will then respond or reply: "put *Responsio ad objectiones,* rather than *Solutiones objectionum,*" he tells Mersenne, "to allow the reader to judge whether or not my responses contain the solutions" (*Oeuvres III,* 340, my translation). One could as well imagine the initial objections being called "responses" to the *Meditations* and Descartes's replies being called "objections." A set of responses, Derrida comments, is also what the 1638 letter to an unknown addressee presents itself as, but within it there uncannily arises, as we have seen, "the question of the response of the automaton, or of the animal as automatic responder [*répondeur automatique,* also "answering machine"] and therefore without response" (*Animal,* 85).

Any reading functions automatically, of course, as a matter of course, by definition, like a mechanical semantic reaction or echo. As we read, we produce a somehow visual, somehow aural, somehow cerebral, eventually perhaps intellectual, repetition of what we read. It is in being tethered to that physio-logical necessity—hence a form of automatism—that reading responds to the call of the text, well before the response we call commentary, exegesis, debate. Descartes repeats that he dislikes the "business of writing books" (*Writing I,* 142), that he has "never had an inclination to produce books,"[16] suggesting that he would prefer not to write and so not to invite or incite such responses. He would prefer to think or sing to himself in autoaffective response or autoresponsive affect. That is how he would protect himself from any possibility of receiving a response tinged with a reaction, from the hint of automatism that he automatically is inclined to manufacture, it seems, whenever it is a question of an animal speaking. Fortunately for us, we haven't had to take him too seriously. Something like reflex, impulse, instinct, or drive—and Freud and Lacan both should have their word to say about the remainder or not, lack or not, of animal in each of those—took over, so that he did indeed write and publish books, even in the face of the Index and Inquisition. Once he gave in to writing, from the invited objections of 1641 all the way down to the present, the answering machine—one that "Descartes could not have imagined in [its] refinements, capacity and complexity, all the powers of reaction-response that today we can, and tomorrow should be better and better able to, attribute to machines, and to another concept of the machine" (*Animal*, 84)—even one with a parrotlike synthesized human voice repeating that the machine is on, is on.

VI. Descartes perhaps steps back from the absolute material falsity of such things as light and colors, sounds, smells, tastes, heat and cold, from their being known, as he maintained in the Second and Third Meditations, "only in a very confused and obscure way." He perhaps undertakes to revise their lack of clarity and distinctness when, in the Sixth Meditation, returning to material things, he accepts that in respect of them he is "taught something by nature," that he may in fact have been "in the habit of misusing the order of nature." For example, still on the matter of heat and cold, he understands that because "I feel heat when I go near a fire and pain when I go too near . . . there is simply reason to suppose that there is something in the fire . . . which produces in us the feelings of heat or pain" (*Writings II,* 57). As I have just suggested, he understands that perhaps thanks to Galileo, Giordano Bruno, and others, and their experience of the Most Eminent Cardinals, or perhaps because he is confident that he can control the ardor of his Santpoort hearth, his *poêle,* and his proximity to it. From the beginning, from 1619 at least, he is next to it: "While I was returning to the army from the coronation of the emperor, the onset of winter detained me in quarters where . . . I stayed all day shut up alone in a stove-heated room (comme je retournais du couronnement de l'empereur vers l'armée, le commencement de l'hiver m'arrêta en un quartier où . . . je demeurais tout le jour enfermé seul dans un poêle; *Writings I,* 116; *Oeuvres VI,* 11). It begins, as we have already seen, at that moment near Ulm that Baillet has localized for us in the biography. He is still by the fire some twenty years later, in the *Meditations,* in reality, in his dressing gown and in his dreams: "I am here, sitting by the fire, wearing a winter dressing-gown. . . . How often, asleep at night, am I convinced of such familiar events—that I am here in my dressing-gown, sitting by the fire—when in fact I am lying undressed in bed" (*Writings II,* 13). He comes closer to it, perhaps close to pain, stretching his hand toward it, if not holding it to it, in order to put his wax where he can watch it melt: "Even as I speak, I put the wax by the fire, and look" (20). Next day, yesterday's wax, restoked fire, still sitting by it, still feeling its heat, although he doesn't tell us what he is wearing (26, 29). Finally, sixth time around, flames and stars in his eyes and the idea, again, of getting so close that it hurts (57).

Fire, as we know from the *Treatise on Man*, is the *anima* or animal itself, the "very lively and pure flame, which is called the *animal spirits*" (*Writings I*, 100). Fire is the privileged example for describing bodily reaction, and the one that is kindled near a hand or foot operates in circular response, reaction or reciprocity with the one in the pineal gland, as graphically shown by the illustrations in the *Treatise*.[17] Of water nothing is said in the *Meditations*, but in the *Treatise* it serves, however paradoxically, to illustrate the means by which that flame of the animal spirits brings about the articulations of the body: "Indeed, one may compare the nerves of the machine I am describing with the pipes in the works of these fountains, its muscles and tendons with the various devices and springs which serve to set them in motion, *its animal spirits with the water which drives them*, the heart with the source of the water, and the cavities of the brain with the storage tanks" (*Writings I*, 101, emphasis added). So goes the elaborate analogy of the royal garden fountains mentioned earlier, with a whole aquatic carnival between pineal gland and external heat source, complete with visitors entering the grottos, a bathing Diana, Neptune, water-spewing sea monsters, and, finally, the rational soul residing there "like the fountain-keeper" (101). A veritable mechanics of the fluid once his imagination gets going, with its mask on, of course, to explain how burning spirits breathe life into limb in the body of these fictitious mechanical men designed to explain the workings of the real men whom "I did not yet have sufficient knowledge to speak of . . . in the same manner as I did of other things" (*Writings I*, 134). Once again, the animal body functions as if in a proliferating call and response between that animal and mechanical other, between reality and fiction, and now also between fire and water.

Even at the pineal seat of things, therefore, we find the expected and now familiar opposition between animal spirit and a type of clockwork automatism, an automatism that is figured in this instance by the Promethean technology deriving from fire, namely, the spontaneous combustion of "one of those fires without light . . . whose nature I understood to be no different from that of the fire which heats hay when it has been stored before it is dry" (134). In addition to that, however, we now also encounter an opposition between such an originary technicity of combustion and one of water. For, as the royal gardens fountains analogy demonstrates, in water more than in fire, there appears to inhere a form of mechanical immediacy, a turbine instantaneity designed to allegorize corporeal articulation. It is in, or out of, such firewater, or waterfire, that life animates, or inanimates, itself.

In Amsterdam, Santpoort, and roundabout, we imagine, there was altogether too much water. No doubt it began as rising damp in the walls that invaded his dressing gown the moment he divested himself of it to slip between the sheets. No doubt he fell asleep to dream of it constantly threatening to drown the hats and coats, natural and automatic alike, out in the street. All through his meditating, Descartes is not about to get dressed to go out and experience it. As if there were always too little fire, far too much water—stay in the dressing gown close to the fire, you're not in the Indre-et-Loire any more. For well nigh two score years, he has been trying to stay dry, since November 1619 striving to stem the onslaught of cold and damp. The room he takes refuge in, in 1619 as in 1637, or thereabouts, to think, write, stay warm, and dream of staying warm, is described by means of an archaic word, such that the word for a stove comes to represent a stove-heated room. *Un poêle* is Descartes's own word, at that time written *poësle* in French, and it appears to have been in usage with that sense of a warm room just long enough for him to use and enjoy it.[18] We can almost read it as a metonymy of his own invention, a case of the fire spreading from its container—already somewhat more artisanal or technological than a simple hearth—extending to engulf the room. Or else we can understand the word itself to have been placed near the fire of signifying drift, melting and extension, sufficient to engulf the room as a liquid, like water, allowing him to relax and meditate as if in a warm bath, coddled by a gaseous or spiritous amnion. In the *poêle* metonymized as heated room, we could perhaps also trace a movement from animal reaction to ratiocinative response, from perceptual feedback to conceptual apparatus, as his heated body cedes the initiative to the space of more or less ardent meditation. At one pole or the other of that movement, we would hope to find life, both inanimate and animal life, life animated by the technological spontaneities that are fire and water as much as in a hand extended to or withdrawing from a hot stove. From one *poêle* to the other there would be inanimate life, beginning in the very homonymic repetition as metonymic displacement itself, *poêle* the stove, *poêle* the room, in the shift from *poêle* to *poêle*, in the automatism of that differential iterability without which there is no movement at all, no automotion, no autoregulation, no auto(re)generation. Hear it, *poêle poêle*, or *poësle poësle*, like *whoit whoit* or *what cheer what cheer*, like a song uttered by one of the mechanical parrots Descartes conjures for himself for company and for warmth inside his stove-room, and imagine him projecting it singingly in front of him as he advances dressing-gowned toward the stove that is the whole stage of his *discursive* and *meditative* life; *poêle poêle*, we hear him sing also as he advances toward the bed that is the whole stage of his oneiric life, finally unmasked except for that very song, *poêle poêle*, how often, asleep at night, is he convinced of such familiar events—that he is there in his dressing gown sitting by the fire—when in fact he is lying undressed in bed ("cum tamen positis vestibus jaceo inter strata; quoy que ie fusse tout nud dedans mon lict"), the First Meditation affirms (*Writings II*, 13; *Oeuvres VII*, 19; *Oeuvres IX*, 14), clothes shed, which is to say *à poil* in French, stark naked, naked as wax or as a jay bird and dreaming of the body he doesn't know he has, singing still as he dreams of sitting by his stove, hearing the same song he has now recorded for himself, as authentic in his voice box, his memory, and his dreams, his real live song of life playing over and over within him or without him.

ACKNOWLEDGMENTS

I view this book as the culmination of work and thinking that began more than thirty years ago. It was then, and there, in Townsville, Australia, that my naive attempts to theorize what I would come to call "prosthesis" met their match in the whiplash wit, intellectual trenchancy, and inimitable Liverpudlianicity of Alec McHoul. It is he whom I wish to acknowledge first of all, to thank him for the example of academic labor that he set for me in those days and for the encouragement he has continued to provide, however silently, up to the present.

More recently, I have benefited from the comments and responses of friends and interlocutors who have heard or read various versions of parts of this book in different venues since 2007. I name and thank here the organizers who invited me to present at a series of conferences (if some of them, mentioned here, are omitted later in a different context, or vice versa, I am nevertheless doubly grateful): Marta Segarra, Hélène Cixous/Croire rêver/Arts de pensée, Université de Paris; Matheson Russell, Australasian Society for Continental Philosophy, University of Auckland; Laurent Milesi, Zoontotechnics, twentieth anniversary conference of the Center for Critical and Cultural Theory, Cardiff University; Dawne Mc-Cance, *Mosaic* Freud after Derrida Conference, University of Manitoba; Chelsea Pomponio, Graduate Romanic Student Association Conference, University of Pennsylvania; and Steven Miller, American Comparative Literature Association Conference, Long Beach. Similarly, the friends and colleagues who invited me to try out these ideas at their institutions included Dragan Kujundžić, University of Florida; Martin McQuillan, London Graduate School; Erich Hörl, Ruhr Universität Bochum; Elissa Marder, Emory University; and Peter Connor, Barnard College. Finally, those who invited me to contribute to special issues of journals or to books were Peggy Kamuf, Anne Berger, John Drabinski, and Marie-Luise Angerer.

I am most grateful to Marguerite Derrida for permission to translate and publish extracts from Jacques Derrida's 1974 seminar "La vie la mort"; to Hélène Cixous for photographs of the Derrida "Un Ver à soie" manuscript; and to Allison Hunter, who kindly worked with me to produce the mark-ups of "Wheelbarrow Angel" and "Red Wheelbarrow."

I am similarly indebted to Posthumanities series editor Cary Wolfe for his unfailing support of my work; to Doug Armato for the warm reception he gives to the series and its authors at Minnesota; and to Nicholas Royle for his generous and perceptive reading of the book. I also wish to thank the entire editorial and production team at Minnesota, beginning with Erin Warholm-Wohlenhaus, and not least Holly Monteith, copy editor supreme.

Since 2013, numerous colleagues have made me feel at home at Brown University. These acknowledgments would not be complete without my naming one of them, Thangam Ravindranathan, whose strong support and solicitous friendship I find exemplary and appreciate enormously.

And then again, thanks to my collaborators in the Derrida Seminars Translation Project, who make every first week of July a delightful anticipation: Peggy Kamuf, Geoff Bennington, Michael Naas, Pascale-Anne Brault, Elizabeth Rottenberg, Kir Kuiken, Katie Chenoweth, Ellen Burt, and student participants, who know who they are.

NOTES

PREFACE

1 For example, from *a*1631, "God hath given our Zeal . . . a new inanimation by this fire of tribulation." John Donne, *The Sermons of John Donne*, ed. George R. Potter and Evelyn M. Simpson (Berkeley: University of California Press, 1953), 4:362. Qtd. in *OED*.

2 Michel Foucault, *The Order of Things: An Archeology of the Human Sciences* (New York: Vintage, 1994).

3 Jacques Derrida, "Nietzsche and the Machine," in *Negotiations*, trans. Elizabeth Rottenberg (Stanford, Calif.: Stanford University Press, 2002), 241.

4 Gilles Deleuze and Félix Guattari, *A Thousand Plateaus: Capitalism and Schizophrenia*, trans. Brian Massumi (Minneapolis: University of Minnesota Press, 1987), 256. Further references included in text are preceded by the mention *Plateaus*. The translation is sometimes slightly modified (cf. *Mille Plateaux* [Paris: Minuit, 1980]), and, for reference, the French *Plateaux* followed by a page number is sometimes added.

5 See David Wills, *Prosthesis* (Stanford, Calif.: Stanford University Press, 1995); Wills, *Dorsality: Thinking Back through Technology and Politics* (Minneapolis: University of Minnesota Press, 2008). The first volume seeks to transform a thematics of artificiality, or an autobiographical fiction, or both, into a performative or operative lever that disrupts the contrived uniformity of organicist presumptions concerning writing; the second seeks to counter instrumentalist presumptions concerning human or animal technology by developing an ethical and political nexus that derives from recognition of what is *behind* or *back*.

INTRODUCTION

1 François Jacob, *The Logic of Life: A History of Heredity*, trans. Betty E. Spillmann (Princeton, N.J.: Princeton University Press, 1993). The original title of the translation was *The Logic of Living Systems* (London: Allen Lane, 1973), which is closer to the nominal participle of the French. Cf. *La logique du vivant: Une histoire de l'hérédité* (Paris: Gallimard, 1970).

Jacob's book is in dialogue with two notable prior publications, more or less explicitly with Foucault's *Les Mots et les choses* (*The Order of Things*; Paris: Gallimard, 1966), at least when it comes to describing the principles of similitude in force during the preclassical period (cf. Jacob, *Logic*, 20–27), and implicitly with Georges Canguilhem's *La connaissance de la vie* (Paris: Vrin, 1955; revised and expanded in 1965), translated by Stefanos Geroulanos and Daniela Ginsburg as *The Knowledge of Life* (New York: Fordham University Press, 2008). Further references to Jacob's book included in text are preceded by the mention *Logic*. The translation is sometimes slightly modified, in which case the French reference (*Logique*) is added.

2 For a recent reassessment, written for a lay readership, see Eve Jablonka and Marion J. Lamb, *Evolution in Four Dimensions: Genetic, Epigenetic, Behavioral, and Symbolic Variation in the History of Life* (Cambridge, Mass.: MIT Press, 2006). As the title and subtitle indicate, the authors argue from current knowledge "that a whole battery of sophisticated mechanisms is needed to maintain the structure of DNA and the fidelity of its replication. Stability lies in the system as a whole, not in the gene" (7). They decline to define life ("the 'definition of life' issue is a really messy subject" [40]) and allow that "maybe we should be talking about different manifestations of life, rather than whether there is a clear distinction between life and nonlife" (41). In other respects, for example, concerning the major lexicon and semantics of code, transcription and translation, copying and error, their language refers less deterministically than Jacob's to the genetic program or text, but the model remains to a great extent intact.

3 Jean-Baptiste de Lamarck, *La Flore française* (Paris: Imprimerie Royale, 1778), 1:1–3, my translation.

4 Jacques Derrida, *The Beast and the Sovereign*, vol. 1, trans. Geoffrey Bennington (Chicago: University of Chicago Press, 2009), 221. Further references included in text are preceded by the mention *B&S I*.

5 Jacques Derrida, "La vie la mort," typescript, Box 12, Folder 11, Special Collections and Archives, Critical Theory Archive, Library of the University of California, Irvine. The seminar consists of fourteen sessions, with discussion of Jacob being found principally in sessions 4–6 and constituting the sole topic of session 6. Further references included in text are preceded by the mention "Life Death" followed by a session number and typescript page number. Quotations are by kind permission of Marguerite Derrida. All translations are mine.

6 Cf. *B&S I* and Jacques Derrida, *The Beast and the Sovereign*, vol. 2, trans. Geoffrey Bennington (Chicago: University of Chicago Press, 2011; further

references included in text are preceded by the mention *B&S II*), and Jacques Derrida, *The Post Card: From Socrates to Freud and Beyond*, trans. Alan Bass (Chicago: University of Chicago Press, 1987), 259; further references included in text are preceded by the mention *Post Card*. The seminar in question will have begun in fall 1974, soon after Derrida published *Glas* (Paris: Galilée, 1974). Session 12 was published in *Études freudiennes* 13–14 (1978) and sessions 11–14 as "To Speculate—on 'Freud,'" in *Post Card*.

7 Jacques Derrida, "White Mythology," in *Margins of Philosophy*, trans. Alan Bass (Chicago: University of Chicago Press, 1982), 261. See Richard Doyle's cogent analysis of the rhetoric and narratology of scientific discourse on life in his *On Beyond Living: Rhetorical Transformations of the Life Sciences* (Stanford, Calif.: Stanford University Press, 1997). According to Doyle, "scientific writing, in its habit for narrative and linear temporality ... organizes and unifies the organism and thus highlights or demarcates the living according to the temporality of signification, which itself must be hidden for the narrative to function" (82). The pages that follow are greatly indebted to Doyle's readings of Jacob and Monod, which reinforce, and at points intersect with, the emphases of Derrida's seminar.

8 Jacques Derrida and Élisabeth Roudinesco, *For What Tomorrow ... : A Dialogue*, trans. Jeff Fort (Stanford, Calif.: Stanford University Press, 2004), 21.

9 Derrida's argument in this section of the seminar develops out of an analysis of the first line of the Francis Ponge poem "Fable"—"Par le mot *par* commence donc ce texte." The analysis is developed in "Psyche: Inventions of the Other," trans. Catherine Porter, in *Psyche: Inventions of the Other*, ed. Peggy Kamuf and Elizabeth Rottenberg (Stanford, Calif.: Stanford University Press, 2007), 1:7–19.

10 See, in his *On Beyond Living*, Doyle's excellent interpretation of Jacob's and Monod's performance of the definition of "gene," such that the DNA message is able "to both instruct (send a message, specify) and construct (define structure) ... to announce genetic commands and to execute them" (73), analyzed against the background of Jacob's autobiographical "re-creat[ion] of the universe" both in his childhood bedroom and in the Pasteur institute (68–69).

11 Denis Diderot and Jean d'Alembert, eds., *Encyclopédie, Dictionnaire raisonné des sciences, des arts et des métiers, par une Société de Gens de lettres*, vol. 17 (1765), 259 (my translation), http://artflsrvo2.uchicago.edu /cgi-bin/philologic/getobject.pl?c.16:381.encyclopedie0513.

12 See, e.g., Jablonka and Lamb, *Evolution in Four Dimensions*, 13, 106–7, 360–61.

13 Jacob dates the first usage to a 1712 memoir by Réaumur. The *OED* gives 1728 as the date of the first comparable English usage. For the *Encyclopédie* definition, see http://artflsrv02.uchicago.edu/cgi-bin/philologic /getobject.pl?c.13:462.encyclopedie0513.

14 See again my *Prosthesis* for extensive development of these ideas. There is an uncanny historical coincidence between the first use of *prosthesis* in English (1553) and the first use, sometime prior to 1555, of *inanimate*. The *OED* quotes from a 1570 history by John Foxe that attributes the usage to preacher and martyr John Bradford (1510–55): "Shall we see sacrifice and Gods seruice done to an inanimate creature and be mumme?" One could also, undoubtedly, make something of the parallel between Bradford, a Protestant convert imprisoned after the death of Edward VI and eventually burned, and Thomas Wilson, a Ramist rhetorician who fled to Rome at the same time and, though briefly imprisoned there, survived the reign of Mary. For discussion of Wilson and of the etymology of *prosthesis*, see *Prosthesis*, 218–32.

15 Canguilhem, *Knowledge of Life,* 60.

16 The sparse references to, and acknowledgment of, secondary materials on Deleuze to be found in the following pages do great injustice to many commentators thanks to whom the philosopher's work, both in his own name and in collaboration with Guattari, lives on. But I would be egregiously remiss not to express my gratitude to various readers of Deleuze who have enabled and complemented my living and thinking for almost four decades now, beginning with two Antipodeans with whom friendship was firmly sealed in Paris in 1976, Paul Patton and Rosina Braidotti, and later also Elizabeth Grosz and Meaghan Morris, as well as those whom I have since come to know, appreciate, and respect in the United States: Charles Stivale, John Protevi, Gregg Lambert, Greg Flaxman, Dan Smith, and Brian Massumi. Beyond that, it is Branka Arsić who knows best all things; this book was written in the company of her forthcoming *Bird Relics: Grief and Vitalism in Thoreau* (Cambridge, Mass.: Harvard University Press, 2016).

17 See Keith Ansell Pearson's excellent discussion of differences between Geoffroy and Cuvier in his *Germinal Life: The Difference and Repetition of Deleuze* (London: Routledge, 1999), 159–61.

18 See also *Plateaus,* 69: "*The plane of consistency knows nothing of differences in level, orders of magnitude, and distances. It knows nothing of the difference between the artificial and the natural.*" Ansell-Pearson's comprehensive analysis of nonorganic life, in particular in relation to Bergson, is again invaluable here. See *Germinal Life,* 42–58ff.

19 Brian Massumi's translation makes explicit the double sense of *plan* in

the first case but not in the second two (cf. *Plateaux,* 325–26, 311–13, 456). In *What Is Philosophy?*, the plane of immanence is defined successively as breath, always fractal unlimited absolute, abstract machine, horizon of events, reservoir or reserve, indivisible milieu, like a desert. But perhaps the most evocative description uses the word *feuilleté,* which gets translated as "interleaved." A little later, it will be not only *feuilleté* but *troué* (holed). Feuilleté is most often used to describe puff or flaky pastry, as in a "thousand-leaved" Napoleon *(millefeuille).* See Gilles Deleuze and Félix Guattari, *What Is Philosophy?*, trans. Hugh Tomlinson and Graham Burchell (New York: Columbia University Press, 1994), 36, 50, 51; cf. *Qu'est-ce que la philosophie* (Paris: Minuit, 1991), 39, 51, 52.

20 Gilles Deleuze, *Pure Immanence: Essays on a Life,* trans. Anne Boyman (New York: Zone Books, 2001), 25–33.

21 See in this regard Deleuze's appendix to *The Logic of Sense* on the crack-up in Zola: "Heredity is not what passes through the crack [*fêlure*], it is the crack itself. . . . Everything rests on the paradox, that is the confusion of this heredity with its vehicle or means, or the confusion of what is transmitted with its transmission . . . nothing other than itself." Gilles Deleuze, *The Logic of Sense,* trans. Mark Lester with Charles Stivale (New York: Columbia University Press, 1990), 321–22, and the ensuing discussion of the instinct for death and the machine, which points in the direction of our chapter 2. One could also evoke Canguilhem here: "Too often, scientists hold the laws of nature to be essentially invariant. . . . From this perspective, the singular—that is, the divergence, the variation—appears to be a failure, a defect, an impurity. The singular is thus always irregular, but that is at the same time perfectly absurd, for no one can understand how a law whose reality is guaranteed by its invariance or self-identity could be at once verified by diverse examples and powerless to reduce their variety, that is, their infidelity." Canguilhem, *Knowledge of Life,* 123. See also his *The Normal and the Pathological,* trans. Caroline R. Fawcett and Robert S. Cohen (New York: Zone Books, 1991), 275–87.

22 Cf. Gilles Deleuze and Félix Guattari, *Anti-Oedipus: Capitalism and Schizophrenia,* trans. Robert Hurley and Mark Seem (New York: Penguin Modern Classics, 2009). Paul Patton has gone a long way toward clarifying the sense of Deleuze and Guattari's machines, particularly as they relate to their conceptual apparatus in general. See, e.g., Patton, *Deleuze and the Political* (London: Routledge, 2000), 1–2, and Patton, *Deleuzian Concepts* (Stanford, Calif.: Stanford University Press, 2010), 32–34.

23 Ansell-Pearson, *Germinal Life,* 142.

24 See also Gilles Deleuze, *Spinoza: Practical Philosophy,* trans. Robert Hurley

(San Francisco: City Lights Books, 1988), whose final chapter (122–30) repeats many of the formulations of *A Thousand Plateaus.*

25 This was the systematic argument of my *Dorsality* (see esp. 5–9). See also Ansell-Pearson, *Germinal Life,* 140–43.

26 Jacques Derrida and Jean Birnbaum, *Learning to Live Finally: The Last Interview,* trans. Pascale-Anne Brault and Michael Naas (New York: Melville House, 2007): "This surviving is life beyond life, life more than death, and my discourse is not a discourse of death but, on the contrary, the affirmation of a living being who prefers living and thus surviving to death, because survival is not simply that which remains but the most intense life possible" (50–51).

27 Cf. Gilles Deleuze, *Cinema 1: The Movement-Image,* trans. Hugh Tomlinson and Barbara Habberjam (Minneapolis: University of Minnesota Press, 1986), and *Cinema 2: The Time-Image,* trans. Hugh Tomlinson and Robert Galeta (Minneapolis: University of Minnesota Press, 1989). Further reference to these volumes included in text are preceded by the mention *Cinema 1* or *Cinema 2.*

28 Walter Benjamin, "The Task of the Translator," in *Selected Writings, vol. 1, 1913–1926,* ed. Marcus Bullock and Michael W. Jennings (Cambridge, Mass.: Harvard University Press, 1996), 254. Further references to volume 1 included in text are preceded by the mention *Writings 1.*

1. AUTOMATIC LIFE, SO LIFE

1 Jean-Luc Nancy, "Larvatus pro deo," trans. Daniel A. Brewer, *Glyph* 2 (1977): 14; cf. Nancy, *Ego Sum* (Paris: Aubier-Flammarion, 1979), 63.

2 René Descartes, *The Philosophical Writings of Descartes, Vol. I.* Trans. John Cottingham, Robert Stoothoff, and Dugald Murdoch (Cambridge: Cambridge University Press, 1985), 2 (further references to this edition are included in text, preceded by the mention *Writings I*); cf. *Oeuvres de Descartes,* ed. Charles Adam and Paul Tannery (Paris: Vrin, 1996), 10:213. Further references to the eleven volumes of the *Oeuvres* included in text are preceded by the mention *Oeuvres* and followed by a volume number.

3 For details regarding the state and history of the manuscript, see Henri Gouhier, *Les Premières pensées de Descartes* (Paris: Vrin, 1958), 11–16, 66.

4 Gilles Deleuze, *Difference and Repetition,* trans. Paul Patton (New York: Columbia University Press, 1994), 110. For Deleuze, though, there is no metamorphosis from larval self to *ego cogito*: "it is not even clear that thought . . . may be related to a substantial, completed and well-constituted subject, such as the Cartesian Cogito: thought is, rather, one of those

terrible movements which can be sustained only under the conditions of a larval subject" (118). I return to something of a zoological larval self in chapter 2. I am grateful to Laurent Milesi for his insights regarding the etymology of *larva*.

5 Charles Darwin, "Self-Attention, Shame, Shyness-Modesty: Blushing," chapter 13 in *The Expression of the Emotions in Man and Animals* (New York: D. Appleton, 1899), 309.

6 For relations between age and blushing, see Stephanie A. Shields, Mary E. Mallory, and Angela Simon, "The Experience and Symptoms of Blushing as a Function of Age and Reported Frequency of Blushing," *Journal of Nonverbal Behavior* 14, no. 3 (1990): 171–87.

7 Cf. Bernard Stiegler, *Technics and Time: Vol. 1. The Fault of Epimetheus*, trans. Richard Beardsworth and George Collins (Stanford, Calif.: Stanford University Press, 1998).

8 Jacques Derrida, *The Animal That Therefore I Am*, trans. David Wills (New York: Fordham University Press, 2008), 5. Further references included in text are preceded by the mention *Animal*. In Elissa Marder's telling analysis of the Abu Ghraib scandal, shame "as an anxiety concerning the fear of unexpected exposure" comes to "have a privileged relation to the medium of photography. . . . Shame is photographic." Marder, *The Mother in the Age of Mechanical Reproduction* (New York: Fordham University Press, 2012), 96. See also her "La Hontise: Confusion, Contagion, Revenants," in *Lire, écrire la honte*, ed. Bruno Chaouat, 89–104 (Lyon, France: Presses Universitaires de Lyon, 2007).

9 See Ray Crozier, "The Puzzle of Blushing," *The Psychologist* 23, no. 5 (2010): "Blushing is a panhuman experience but its visibility depends on facial complexion. . . . The colour red [i]s associated with shame in 78 languages, predominantly in populations with fair complexions. In populations with relatively darker complexions shame was sometimes associated with black, perhaps referring to darkening of the skin due to increased blood flow in the blush region" (393).

10 See Gouhier, disputing Adam: "Why would these schoolboys be 'ashamed' to mount the stage after being led there by their teachers? It is not a matter of 'shame' but of timidity, and professional actors are susceptible to 'fright' even when they are quite accustomed to the stage." Gouhier, *Premières pensées*, 68n35, my translation. See also Nancy, "Larvatus pro deo," on Descartes's relation to baroque theater and the idea that "Descartes's thought is indeed the most exemplary form of *baroque thought*" (32, translation modified; cf. Nancy, *Ego sum*, 91).

11 See my *Prosthesis*, 12–26, 214, 225–26.

12 Nancy, "Larvatus pro deo," 33; cf. Nancy, *Ego sum*, 92.

13 Cf. Antoine Baillet, *La Vie de M. Des-Cartes. 1ère Partie* (Paris: Daniel Horthemels, 1691), 81–86.

14 See Gouhier, *Premières pensées,* 137–38ff.

15 On all these matters, see, e.g., Dalia Judovitz, *Subjectivity and Representation in Descartes* (Cambridge: Cambridge University Press, 1988), 87ff.; Nancy, *Ego sum,* 63–64; Jean-Luc Marion, *Cartesian Questions* (Chicago: University of Chicago Press, 1999), 1–19.

16 See Baillet, *Vie de Des-Cartes,* 80–86.

17 See the introduction to this volume.

18 See my *Dorsality.*

19 "On Language as Such and the Language of Man," *Writings 1,* 64, 73. Benjamin doesn't make clear what constitutes "inanimate nature": non-animal nature (e.g., plants), rocks, artifice, or all three categories.

20 On the ethics of mourning, see again Arsić, *Bird Relics.*

2. ORDER CATASTROPHICALLY UNKNOWN

1 Translations from *Post Card,* as in this case, are often modified.

2 Jacques Derrida, "To Speculate—on 'Freud,'" in *Post Card,* 259. Further references to this essay from *Post Card* will be included in text preceded by the mention "Speculate."

3 Sigmund Freud, "On Narcissism: An Introduction," in *Standard Edition of the Complete Psychological Works of Sigmund Freud,* ed. James Strachey (London: Hogarth Press, 1966), XIV:78–79. Further references to the *Standard Edition* included in text are preceded by the mention *SE* and a volume number.

4 The canonical discussion of this question is to be found in Frank J. Sulloway, *Freud, Biologist of the Mind* (New York: Basic Books, 1979). Sulloway is attentive both to Freud the scientist versus Freud the psychological theorist, and to Freud the biophysicalist versus Freud the Lamarckian evolutionist.

5 Cf. Sigmund Freud, *Civilization and Its Discontents, SE XXI,* 64. Elissa Marder brilliantly reads the ego's "uncanny" relation to itself as a function of the role played by anxiety in Freud's adjustment of his topographical model. Marder, *Mother in the Age of Mechanical Reproduction,* esp. 83–86.

6 See "Editor's Introduction" to "Papers on Metapsychology," *SE XIV,* 106.

7 "Selves are larval subjects; the world of passive syntheses constitutes the system of the self . . . the system of a dissolved self." Deleuze, *Difference and Repetition,* 78.

8 Cf. Jacques Derrida, "The Law of Genre," trans. Avital Ronell, and "Living On," trans. James Hulbert, in *Parages,* ed. John P. Leavey (Stanford, Calif.: Stanford University Press, 2011), esp. 234–36, 124–27.

9 I have discussed this gesture extensively in my *Prosthesis*, 93–104, with emphasis on the rhetorico-prosthetic redirection of the superego.

10 See Derrida's analysis of this theoretical washing of the hands by Freud in "Speculate," 344.

11 Jacques Derrida, "Freud and the Scene of Writing," in *Writing and Difference*, trans. Alan Bass (Chicago: University of Chicago Press, 1978), esp. 202, 227–28. See also "Speculate": "there is never repetition *itself*. Sometimes repetition, classically repeats something that precedes it. . . . But sometimes, according to a logic that is other, and non-classical, repetition is 'original,' through an unlimited propagation of itself, a general deconstruction: not only of the entire classical ontology of repetition, along with all the distinctions recalled a moment ago, but also of the entire psychic construction" (351–52).

12 Quoted in Sulloway, *Freud, Biologist*, 394.

3. THE BLUSHING MACHINE

1 Martin Heidegger, *The Fundamental Concepts of Metaphysics: World, Finitude, Solitude*, trans. William McNeill and Nicholas Walker (Bloomington: Indiana University Press, 1995), 253.

2 Cf. Marie-Louise Mallet, ed., *La démocratie à venir: autour de Jacques Derrida* (Paris: Galilée, 2004).

3 See Marie-Louise Mallet's foreword to *Animal*, xi–xii. Mallet's preface was written before the decision was made to publish the seminars, in particular, that of 2002–3, which contained extensive development of the Heidegger material. See *B&S II* and my following discussion.

4 See Jacques Derrida, *Signéponge/Signsponge*, trans. Richard Rand (New York: Columbia University Press, 1984), 22, and Derrida, *Limited Inc.*, ed. Gerald Graff (Evanston, Ill.: Northwestern University Press, 1988), 18.

5 See also *Memoires*: "In calling or naming someone while he is alive, we know that his name can survive him and already survives him; the name begins during his life to get along without him, speaking and bearing his death each time it is pronounced in naming or calling, each time it is inscribed in a list, or a civil registry, or a signature." Jacques Derrida, *Memoires: For Paul de Man*, trans. Cecile Lindsay, Jonathan Culler, Eduardo Cadava, and Peggy Kamuf (New York: Columbia University Press, 1989), 49.

6 See Catherine Malabou and Jacques Derrida, *Counterpath: Traveling with Jacques Derrida*, trans. David Wills (Stanford, Calif.: Stanford University Press, 2004), 5, 21, 269.

7 Derrida, *Writing and Difference*, 227.

8 See Jacques Lacan, "The Function and Field of Speech and Language in Psychoanalysis," in *Ecrits: A Selection,* trans. Alan Sheridan (New York: W. W. Norton, 1977), 85. I take up this point again in chapter 9.

9 Derrida, *Negotiations,* 244.

10 Jacques Derrida, *Rogues: Two Essays on Reason,* trans. Pascale-Anne Brault and Michael Naas (Stanford, Calif.: Stanford University Press, 2004); Derrida and Birnbaum, *Learning to Live Finally.*

11 Derrida and Roudinesco, *For What Tomorrow,* 21.

12 "The choice of the title for this seminar, *La bête et le souverain,* was designed in the first place to keep bringing us back to this first site of decision, as to the immense question of the living" (*B&S I,* 176).

13 See also Jacques Derrida, *Aporias,* trans. Thomas Dutoit (Stanford, Calif.: Stanford University Press, 1993), 35–42.

14 Derrida, *Limited Inc., 7–10ff.*

15 Jacques Derrida, "Typewriter Ribbon: Limited Ink (2)," trans. Peggy Kamuf, in *Without Alibi,* ed. Peggy Kamuf (Stanford, Calif.: Stanford University Press, 2002), 133. Derrida adds the clarification that quasi-machinelike is not the same as machine-like: "This cut is not so much effected by the machine (even though the machine can in fact cut and repeat the cut in its turn) as it is the condition of production for a machine. The machine is *cut* as well as *cutting* with regard to the living present of life or of the living body. The machine is an *effect of the cut* as much as it is a *cause of the cut*" (133). See the whole essay for a recasting of the relations among body, machine, and autobiography that are in question here.

16 Jacques Derrida, "Faith and Knowledge: The Two Sources of 'Religion' at the Limits of Reason Alone," trans. Samuel Weber, in *Religion,* ed. Jacques Derrida and Gianni Vattimo (Stanford, Calif.: Stanford University Press, 1998), 42.

17 Ibid., 45–46.

18 Michael Naas, *Miracle and Machine: Jacques Derrida and the Two Sources of Religion, Science, and the Media* (New York: Fordham University Press, 2012), 86.

19 Ibid.

20 Ibid., 151.

21 Jacques Derrida, *Politics of Friendship,* trans. George Collins (London: Verso, 1997), 76. See also Derrida, *Rogues,* 101–2.

22 Derrida, *Learning to Live,* 25, 32.

23 Cf. Emmanuel Levinas, *Otherwise Than Being or Beyond Essence,* trans. Alphonso Lingis (Boston: Martinus Nijhoff, 1981).

4. LIVING PUNCTUATIONS

1   Apart from what is immanent to the argument itself, my strategy also involves displacing the discussion from a semiotic tradition represented by Jakobson as much as from a semantico-tropological tradition represented by De Man, approaches with which this discussion nevertheless has certain affinities. I take a more marked distance from the new hermeneutics of Quentin Meillassoux's *The Number and the Siren*, trans. Robin Mackay (New York: Sequence Press, 2012). Meillassoux's arithmetical "decipherment" of *Un coup de dés* makes no attempt to quantify or otherwise account for what, as I suggest in what follows, is the single most explicit signifier of Mallarmé's poem, namely, its white space. His omission is all the more striking given that the count on which he bases his decipherment turns, in the final analysis, around usages of the hyphen, requiring him to manipulate the very effects of spacing that he ignores (198–208).

2   Marcel Broodthaers, *Un coup de dés jamais n'abolira le hasard*, print on paper (Brussels: Musées royaux des Beaux-Arts de Belgique). See also Rachel Haidu, *The Absence of Work: Marcel Broodthaers, 1964–1976* (Cambridge, Mass.: MIT Press, 2010).

3   Cf. Stéphane Mallarmé, "'Quant au livre': *Tout, au monde, existe pour aboutir à un livre*" [Everything in the world exists to end up as a book], in *Œuvres complètes,* ed. Henri Mondor and G. Jean-Aubry (Paris: Gallimard/Pléiade, 1945), 378, my translation. See also Jacques Scherer, *Le "Livre" de Mallarmé* (Paris: Gallimard, 1957).

4   Jean Racine, *Phèdre*, trans. Margaret Rawlings (New York: E. P. Dutton, 1962), 48–49.

5   Stéphane Mallarmé, "Un Coup de dés"/"A Throw of the Dice," in *Collected Poems*, trans. Henry Weinfield (Berkeley: University of California Press, 1994), 142–43.

6   "J'apporte en effet des nouvelles. Les plus surprenantes. Même cas ne se vit encore. On a touché au vers." (Indeed, I bring news. Of the most surprising kind. Such a case was never before seen. The (poetic) line has been interfered with.) Stéphane Mallarmé, "La Musique et les lettres," in *Œuvres complètes,* 643, my translation.

7   Gertrude Stein, "Poetry and Grammar," in *Writings and Lectures 1911–1945,* ed. Patricia Meyerowitz (London: Peter Owen, 1967), 127–28.

8   Hélène Cixous, *First Days of the Year,* trans. Catherine A. F. McGillivray (Minneapolis: University of Minnesota Press, 1998), 9–10, translation modified. Further references will be included in text, preceded where

necessary by the mention *First Days*; translations will often be slightly modified to conform to what I am emphasizing. Cf. *Jours de l'an* (Paris: Des Femmes, 1990), 13–14.

9 Hélène Cixous, *Dedans* (Paris: Des femmes, 1986), 95 (note that in French text, semicolons and colons, like question and exclamation marks, are preceded as well as followed by a blank space); Cixous, *Le Troisième corps* (Paris: Des Femmes, 1970), 60, 78; Cixous, *Angst* (Paris: Des Femmes, 1977), 71, 233, 200, 232; Cixous, *Insister: À Jacques Derrida* (Paris: Galilée, 2006), 13–15.

10 Jacques Derrida, *H.C. for Life, That Is to Say,* trans. Laurent Milesi and Stefan Herbrechter (Stanford, Calif.: Stanford University Press, 2008), 62.

11 Ibid., 68.

12 Ibid., 107–8.

13 Wills, *Dorsality,* 7–9, 46, 158–59, 220–22.

14 Derrida, *H.C. for Life,* 119, 120. For *Belebtheit,* see Sigmund Freud, *Totem and Taboo,* where the word is translated as "the living character" (*SE XIII,* 75) or simply as "life" (91). For a rich and inventive discussion of relations between magic and technology in Derrida, see Naas, *Miracle and Machine.*

15 Hélène Cixous, *Dream I Tell You,* trans. Beverley Brahic (Edinburgh: Edinburgh University Press, 2005), 3–4, quoted in Jacques Derrida, *Geneses, Genealogies, Genres, and Genius,* trans. Beverley Brahic (New York: Columbia University Press, 2006), 34.

16 Ibid., translation modified.

17 I am improvising here on the famous formula for the surrealist image, borrowed from Lautréamont: "beau ... comme la rencontre fortuite sur une table de dissection d'une machine à coudre et d'un parapluie" (beautiful like the chance meeting on a dissecting table of a sewing machine and an umbrella). Isidore Ducasse, comte de Lautréamont, *Les Chants de Maldoror,* in *Oeuvres completes,* ed. Hubert Juin (Paris: Gallimard, 1973), 233–34. Also *Maldoror and the Complete Works of the Comte de Lautréamont,* trans. Alexis Lykiard (Cambridge: Exact Change, 1998), 193. See also André Breton, "Manifeste du surréalisme," in *Manifestes du surréalisme* (Paris: Jean-Jacques Pauvert, 1962), 47, also published as "Manifesto of Surrealism," in *Manifestoes of Surrealism,* trans. Richard Seaver and Helen R. Lane (Ann Arbor: University of Michigan Press, 1972), 38; and André Breton, *L'Amour fou,* in *Oeuvres completes II,* ed. Marguerite Bonnet (Paris: Gallimard/Pléiade, 1992), 679, also published as *Mad Love,* trans. Mary Ann Caws (Lincoln: University of Nebraska Press, 1987), 9.

18 See my "Techneology or the Discourse of Speed," in *The Prosthetic Impulse: From a Posthuman Present to a Biocultural Future*, ed. Marquard Smith and Joanne Morra (Cambridge, Mass.: MIT Press, 2005), 257–59.

19 Derrida, *H.C. for Life*, 73, translation modified. Derrida's references to speed are numerous here: "she goes faster than speed" (63), a tautology involves "quasi-infinite acceleration" (72), "infinite speed" (72), "absolute speed, the speed that absolutely economizes on speed" (73). By the same token, he adds, "Let us not act as if, speaking of absolute speed, we could say: I know what speed, the essence of speed, is . . . the question 'what is speed?' and any possible answer to this question . . . arrives at a certain speed. . . . There is no *essence* of speed, nor a metalanguage, nor a theorem for it, outside this differential" (72–73).

20 Paul Celan, "Cello-Einsatz," in *Poems of Paul Celan*, trans. Michael Hamburger (New York: Persea Books, 2002), 236–37.

21 Cf. *Jours de l'an*: "Ce . . . , ce poème, donc" (8); "Ce . . . « poème » donc" (9).

22 Paul Celan, *The Meridian: Final Version-Drafts-Materials*, ed. Bernhard Böschenstein and Heino Schmull, trans. Pierre Joris (Stanford, Calif.: Stanford University Press, 2011). Further references included in text are preceded by the mention *Meridian*.

23 Hélène Cixous, *Insister of Jacques Derrida*, trans. Peggy Kamuf (Stanford, Calif.: Stanford University Press, 2007), 59. Derrida died in October 2004. Further references included in text are preceded by the mention *Insister*.

24 Hélène Cixous and Jacques Derrida, *Veils*, trans. Geoffrey Bennington (Stanford, Calif.: Stanford University Press, 2001): "Before the verdict, my verdict, before, befalling me" is how Derrida's contribution, entitled "A Silkworm of One's Own," opens (21). The word *verdict* resonates both as "sentence" and as "truth-telling," in the sense of "witnessing" as well as making reference to truth and femininity as developed by Derrida through the thematics of veils and sails in his *Spurs*, trans. Barbara Harlow (Chicago: University of Chicago Press, 1981); cf. *Insister*, 57–58.

25 One could insert here the extraordinary commentary by Peter Szendy, *À coup de points: la ponctuation comme expérience* (Paris: Minuit, 2013), concerning Hegel's inanimating lifepoints *(Lebenspunkte)* in that "grand stigmatological fable" (93) that is his *Philosophy of Nature*. See esp. 93–98.

26 *Insister*, 20. This contrast is made most explicit by Derrida in *H.C. for Life*: "It seems at first that for her, and I do mean for her, there is only one side and not two, and this side is that of life. Death, which she knows and understands as well as anyone, is never denied, certainly . . . but it is not a side, it is a nonside. This is why I—and this is probably more than a difference, a big disagreement [*différend*] between us . . . —I, who always feel turned toward death, I am not on her side, while she would

like to turn everything and to make it come round to the side of life" (36). The figure of two sides has been prompted by Cixous's recounting, in a text from 1992, how when she first "saw Jacques Derrida (it had to have been 1962), he was walking along the crest of a mountain . . . the crest was extremely fine, he was walking on its edge [*la cime*]." Hélène Cixous, "Quelle heure est-il ou La porte (celle qu'on ne passe pas)," in Marie-Louise Mallet, ed.), *Le Passage des frontières. Autour du travail de Jacques Derrida,* ed. Marie-Louise Mallet (Paris: Galilée, 1994), 83, my translation.

27 Hélène Cixous, *Portrait of Jacques Derrida as a Young Jewish Saint,* trans. Beverley Bie Brahic (New York: Columbia University Press, 2004), 63.

28 Ibid., 64–65, translation modified.

29 Ibid., 54. Derrida's "Circumfession" reference is from Geoffrey Bennington and Jacques Derrida, *Jacques Derrida,* trans. Geoffrey Bennington (Chicago: University of Chicago Press, 1993), 51.

30 "L'oeuvre pure implique la disparition élocutoire du poëte, qui cede l'initiative aux mots" (The pure work implies the elocutionary disappearance of the poet, who cedes the initiative to words). Mallarmé, "Crise de vers," in *Oeuvres complètes,* 366, my translation.

31 Citations between pages 2 and 13 are from the published version of the address; those from 16–82 are from earlier drafts; and those from 84–213 from source materials and notes.

32 Cf. *Meridian,* 154: "Büchner's hostility to Art—permit the translator of *La Jeune Parque* [Young fate] this confession—I share it."

33 Derrida outlines the complexity of the alterity expressed here: "It is not a matter of speaking time but of . . . *giving* the other its time. . . . It is the time that one must let speak, the time of the other, rather than leaving the other speaking time. It is a matter of letting the time speak, the time of the other in what is most proper to the other, and therefore in what in the other is most other—and which happens, that I let happen, as time of the other, in the present time of "my" poem" (*B&S I,* 234).

34 The final version states, "I am searching for all of this . . . on the map—a child's map I have to confess" (12). The son's globe is mentioned in "Additional Notes to the Text": "I have a son who is five and a half, he is interested, like his father, in the earthly: he owns a globe" (82).

35 Paul Celan, "The Meridian," trans. Jerry Glenn, in Jacques Derrida, *Sovereignties in Question: The Poetics of Paul Celan,* ed. Thomas Dutoit and Outi Pasanen (New York: Fordham University Press, 2005), 179.

36 Celan, *Poems of Paul Celan,* 237.

37 Paul Celan, *Breathturn,* trans. Pierre Joris (Los Angeles, Calif.: Sun and Moon Press, 1995), 184–85.

38 Cf. *Meridian*, editors' preface, xvi. There is also this perhaps loaded choice of the word *concentration*: "The attention the poem tries to pay for everything it encounters... is rather a concentration that remains mindful of all our dates" (9). See also Jacques Derrida, "Shibboleth," in *Sovereignties in Question*, 10, and, on the date in Celan in general, Philippe Lacoue-Labarthe, *Poetry as Experience*, trans. Andrea Tarnowski (Stanford, Calif.: Stanford University Press, 1999).

39 I attempted to address this question, in a different register, vis-à-vis the work of Rimbaud in *Dorsality*, 119–27.

40 Derrida, *Sovereignties in Question*, 6.

41 Ibid., 7. At the time of writing "Shibboleth," Derrida was himself unaware of the reference to Wannsee (cf. *B&S I*, 226).

42 Ibid., 8–9.

43 See also ibid., 128. My version of the note does not show all erasures and rewrites.

44 Werner Hamacher, *Premises*, trans. Peter Fenves (Stanford, Calif.: Stanford University Press, 1996), 368. Hamacher quotes Theodor Adorno, "Cultural Criticism and Society," in *Prisms*, trans. Samuel and Shierry Weber (Cambridge, Mass.: MIT Press, 1981), 34, and Peter Szondi, "Durch die Enge geführt," *Schriften* (Berlin: Suhrkamp, 1978), 2:384.

45 Derrida, *H.C. for Life*, 73.

46 Ibid., 61.

47 Ibid., 70, emphasis added. For a parallel meditation on *pouvoir*, see *B&S II*, 233–35.

5. NAMING THE MECHANICAL ANGEL

1 Derrida, *H.C. for Life*, 73.

2 Ibid., 65.

3 Ibid.

4 In "Des tours de Babel," Derrida refrains from analyzing "On Language as Such" given, in his view, "the overly enigmatic character of that essay, its wealth and its overdeterminations." Jacques Derrida, "Des tours de Babel," trans. Joseph F. Graham, in *Psyche: Inventions of the Other* (Stanford, Calif.: Stanford University Press, 2007), 1:200.

5 The same reproach of a bourgeois view of language is made with respect to the idea "that the word has an accidental relation to its object, that it is a sign for things... agreed by some convention" (*Writings 1*, 69). By the same token, Benjamin distinguishes what he is advancing from the mystical theory that "the word is simply the essence of things" (ibid.).

6 Derrida, *Psyche*, 1:197–98.

7   Cf. Walter Benjamin, "Über Sprache überhaupt und über die Sprache des Menschen," in *Gesammelte Schriften,* vol. 2, ed. Rolf Tiedemann and Hermann Schweppenhäuser (Frankfurt: Suhrkamp, 1991), 150. Further references included in text are preceded by the mention *Gesammelte Schriften.*

8   Carol Jacobs, *In the Language of Walter Benjamin* (Baltimore: The Johns Hopkins University Press, 1999), 91. Also, "from the very early essay 'On Language as Such and on the Language of Man' (1916) a radical concept of translation is already at play *within* the act of naming" (34). The current discussion owes much to Jacobs's careful readings.

9   Walter Benjamin, "On the Concept of History," in Benjamin, *Selected Writings, vol. 4, 1938–1940,* ed. Howard Eiland and Michael W. Jennings (Cambridge, Mass.: Harvard University Press, 2003), 392.

10 Walter Benjamin, *The Arcades Project,* trans. Howard Eiland and Kevin McLaughlin (Cambridge, Mass.: Harvard University Press, 1999), 470; *Gesammelte Schriften,* V, 587.

11 Benjamin, "On the Concept of History," 390–91, emphasis added.

12 Ibid., 395, 396; *Gesammelte Schriften,* I, 701, 702.

13 Ibid., 396.

14 Benjamin, *Arcades Project,* 462.

15 Ibid., 392. The discussion that follows begins in my *Dorsality,* 202–3.

16 See the iconography of the Annunciation in such famous cases as Fra Angelico, 1426, Prado, Madrid, and 1438–45, San Marco, Florence; and Titian, c. 1535, Scuola Grande di San Rocco, Venice, and 1559–64, San Salvador, Venice.

17 Jacobs, *In the Language,* 86.

18 Benjamin, *Arcades Project,* 458, 460, 461.

19 Ibid., 464. Jacobs makes an astute connection with the astrological constellation described in "Doctrine of the Similar," "a constellation that at once, unimaginably, inscribes and is read by the astrologer." Jacobs, *In the Language,* 100.

20 Benjamin, *Arcades Project,* 460.

21 Ibid., 476.

22 Ibid., 458.

23 Ibid., 461.

24 See Jacobs's analysis of reading in "Emergency, Break," a reading that is finally "like the moment of the birth of the human, like the moment of Adam's self-naming both as woman and other, like the gesture of the historical materialist." Jacobs, *In the Language,* 113.

25 See http://www.musee-orsay.fr/index.php?id=851&L=1&tx_commen taire_pi1[showUid]=339.

26   See Salvador Dali, *The Tragic Myth of Millet's Angelus: Paranoiac Critical Interpretation Including the Myth of William Tell*, trans. Eleanor R. Morse (St. Petersburg, Fla.: The Salvador Dali Museum, 1986).
27   See, e.g., J. Hillis Miller, *Poets of Reality* (Cambridge, Mass.: Harvard University Press, 1965), and Charles Altieri, *Painterly Abstraction in Modernist American Poetry* (Cambridge: Cambridge University Press, 1989).
28   Henry M. Sayre, *The Visual Text of William Carlos Williams* (Urbana: University of Illinois Press, 1983), 7.
29   William Carlos Williams, *The Collected Poems of William Carlos Williams, vol. 1, 1909–1939*, ed. A. Walton Litz and Christopher MacGowan (New York: New Directions, 1986), 219.
30   Sayre, *Visual Text*, 65.
31   Ibid., 66.
32   Williams, *Collected Poems*, 229.

6. RAW WAR

1   Cf. Carl Schmitt, *The Concept of the Political*, trans. George Schwab (Chicago: Chicago University Press, 1996), 26–27. Further references to this edition will be included in text, preceded by the mention *Political*. The discussion here pursues questions raised in my *Dorsality* (133–37), in relation to Derrida, *Politics of Friendship*.
2   See Carl Schmitt, *Theory of the Partisan*, trans. G. L. Ulmen (New York: Telos, 2007). Schmitt's endorsement of the "telluric" partisan who defends national soil (16–22) is immediately problematized by "technicization and motorization" (16). Ultimately, these spatial considerations will "flow into the force field of technical-industrial development" (68) to the extent of producing a "cosmopartisan" (80) whose universal, even supra-planetary front calls into question enmity itself (94). See also Derrida's discussion in *Politics of Friendship*, 142–43, as well as his analysis of the "war without war and without front" in his Pantion University address, delivered during the Kosovo War in June 1999: Jacques Derrida, "Unconditionality or Sovereignty: The University at the Frontiers of Europe," trans. Peggy Kamuf, *Oxford Literary Review* 31 (2009): 115–31. The contemporary instance of such a war was launched only fourteen months after that address and remains very much with us, namely, the "state of war [that] exists between the United States of America and any entity determined by the President to have planned, carried out, or otherwise supported the attacks against the United States on September 11, 2001." U.S. Senate Joint Resolution 22/23, September 12–13, 2001. The current "conclusion" of that war, particularly in the form of new interventions in Iraq and Syria

and drone attacks on Pakistan, Yemen, and Somalia, gives rise to the following derision, reported in 2009: "Asked why she's ditching her U.S. boyfriend, a Pakistani woman says, 'He shoots his missile from 30,000 ft.'" Bobby Ghosh and Mark Thompson, *Time,* June 1, 2009. I develop further questions relating to the U.S. targeted killing policy in "Drone Penalty," *Substance* 43, no. 2 (2014): 174–92.

3 Derrida, "Unconditionality or Sovereignty," 121.

4 See Elaine Scarry's extensive discussion of such possibilities in her *The Body in Pain* (Oxford: Oxford University Press, 1985), 81–108.

5 Ernst Jünger, *The Storm of Steel,* trans. Basil Creighton (London: Chatto and Windus, 1929), 1. Creighton's translation, the first to appear in English, is based on the German text revised by Jünger in 1924. Subsequent revisions progressively tempered the language of earlier versions. The current English edition (trans. Michael Hofmann [London: Penguin, 2004]) is from Jünger's 1961 German edition. I prefer to quote from the earlier, more raw version, to which subsequent page references, in text, preceded by the mention *Storm,* refer.

6 James Joyce, *Finnegans Wake* (London: Penguin, 2000), 258. See Jacques Derrida, "Two Words for Joyce," trans. Geoffrey Bennington, in *Post-Structuralist Joyce: Essays from the French,* ed. Derek Attridge and Daniel Ferrer (Cambridge: Cambridge University Press, 1984), and "Reply," trans. Peggy Kamuf, in *The Ear of the Other,* ed. Christie McDonald (Lincoln: University of Nebraska Press, 1985). What follows draws directly on Derrida's reading of Joyce's "he war."

7 *Webster's New World Dictionary,* second college ed. (New York: William Collins/World, 1974), s.v. "meteor."

7. BLOODLESS COUP

1 Georges Bataille, *Le Bleu du ciel,* in *Romans et récits,* ed. Jean-François Louette (Paris: Gallimard/Pléiade, 2004), 201, my translation. Leslie Hill has pointed also to the problematic geographical "sovereignty" of Trier in the demilitarized (1918) then reoccupied (1936) Rhineland. Hill, *Bataille, Klossowski, Blanchot: Writing at the Limit* (Oxford: Oxford University Press, 2001), 82.

2 Bataille, *Bleu du ciel,* 200–201, my translation.

3 Johann Wolfgang von Goethe, *The Sorrows of Young Werther and Novella,* trans. Elizabeth Mayer and Louise Bogan (New York: Random House, 1971), 7.

4 Ibid., 158.

5 Jean-Luc Nancy, "The Heart of Things," trans. Brian Holmes and Rodney

Trumble, in *The Birth to Presence* (Stanford, Calif.: Stanford University Press, 1993), 170. Further references included in text are preceded by the mention "Heart."

6   Jean-Luc Nancy, "Exscription," trans. Katherine Lydon, in *Birth to Presence*, 319. *Sens* translates into English as both "sense" and "meaning." The English translation of this text combines two pieces Nancy wrote eleven years apart and does not include the last sentence cited, which is my translation. I am using the version that goes by the same title in *Une pensée finie* (Paris: Galilée, 1990), 55. Translations, as in this case, are often modified.

7   Jean-Luc Nancy, "Le Partage, l'infini et le jardin," interview with Jean-Baptiste Marongiu, *Libération* 20 (February 2000), my translation.

8   Cf. Jacques Derrida, *Of Grammatology*, trans. Gayatri Spivak (Baltimore: Johns Hopkins University Press, 1976), 287–88. We should note two things here: (1) Nancy's promotion of exscription as a "reaction" against the craze for "writing [*écriture*]" would need to be received as a call for a more rigorous understanding of Derrida's concept of writing: "No thinking about 'writing' has had anything at stake but this: the stake of the *thing* [as exscribed]" ("Heart," 176); (2) the reference to plowing as a relating of language to agriculture not only underlines the originarily technological function of language that is a theme throughout *Of Grammatology*, and of course throughout Derrida's writing in general, but also *reinscribes* that function within Nancy's concept of *exscription*, as I hope to show.

9   See, to begin (for the question pervades and structures much of Nancy's work), *Ego sum*, in particular the final chapter, "Unum quid": "The subject throws itself into this abyss. But *ego* is enunciated there: it is exteriorized there, which does not mean that it brings to the outside the visible face of an invisible interiority. It means literally that *ego* makes or makes itself *exteriority*, spacing of places, gap and strangeness that make up the place, and thus space itself, primary spatiality of a true *tracing* in which, and only in which, *ego* can come about, and trace itself, and think itself" (163). The translation is by Peggy Kamuf in her remarkably astute "Béance," *New Centennial Review* 2, no. 3 (2002): 52.

10   "Le Partage, l'infini et le jardin," my translation.

11   Cf. Deleuze and Guattari, *What Is Philosophy*: "It is the brain that says *I*, but *I* is an other. . . . The search for sensation is fruitless if we go no farther than reactions and the excitations that they prolong. . . . Sensation, then, is on a plane that is different from mechanisms, dynamisms, and finalities. . . . Not every organism has a brain, and not all life is organic, but everywhere there are forces that constitute microbrains or an inorganic life of things" (211–13).

12 Jean-Luc Nancy, *Corpus,* trans. Richard Rand (New York: Fordham University Press, 2008), 71. Further references included in text are preceded where necessary by the mention *Corpus.*

13 See also Derrida: "There is no natural, originary body: technology has not simply added itself, from the outside or after the fact, as a foreign body. Or at least this foreign or dangerous supplement is 'originarily' at work and in place in the supposedly ideal interiority of the 'body and soul.' It is indeed the heart of the heart." Derrida, "The Rhetoric of Drugs," trans. Michael Israel, in *Points . . . Interviews, 1974–1994* (Stanford, Calif.: Stanford University Press, 1995), 244–45.

14 Jean-Luc Nancy, "Shattered Love," trans. Lisa Garbus and Simona Sawhney, in *A Finite Thinking,* ed. Simon Sparks (Stanford, Calif.: Stanford University Press, 2003), 249. Cf. "L'amour en éclats," *Une pensée finie,* 231: "l'amour est le mouvement extrême, au-delà de soi, d'un être s'achevant." Nancy's "definition" is followed, in "Shattered Love," by seven glosses upon it. In the French edition of *Une pensée finie,* "L'amour en éclats" ("Shattered Love") immediately follows "Le coeur des choses" ("The Heart of Things"), which is not found in *A Finite Thinking,* having already been published in *Birth to Presence* (see note 5). Further references to the *Finite Thinking* version of "Shattered Love" appear in text, preceded by the mention "Love."

15 Cf. *Corpus:* "Writing: touching upon extremity" (9).

16 See Denis de Rougemont, *Love in the Western World,* trans. Montgomery Belgion (Princeton, N.J.: Princeton University Press, 1983); Michel Foucault, *The History of Sexuality, Vols. 1–3,* trans. Robert Hurley (New York: Vintage, 1990); Niklas Luhmann, *Love as Passion: The Codification of Intimacy,* trans. J. Gaines and D. L. Jones (Stanford, Calif.: Stanford University Press, 1998); Jean-Luc Marion, *The Erotic Phenomenon,* trans. Stephen E. Lewis (Chicago: University of Chicago Press, 2008). Cf. *Le phénomène érotique* (Paris: Grasset, 2003). Marion's thesis calls for extended discussion of its own. I simply note the following potential elements of such a discussion. (1) On one hand, his "erotic reduction" (*Erotic Phenomenon,* 22) performs an extraordinary subversion of the Cartesian *ego cogitans* in favor of the *ego amans* that Descartes repressed from philosophical thinking (6–7), and on the other, it reinforces the Cartesian divide between human and animal concerning the capacity to love ("in this world only man loves" [7]), expressed more than once as a reductive, humanist anthropocentrism or technophobia: "to give up on even the possibility that someone loves me would be like operating a transcendental castration upon myself, and would bring me down to the rank of an artificial intelligence, a mechanical [*machinal*] calculator

or a demon, in short very likely lower than an animal. . . . [It] implies nothing less than giving up on the human itself" (20–21). (2) That subversion aims foremost at subjectivity in terms that often intersect with those of Nancy—"The erotic reduction renders destitute all identity of self to self" (37)—and consistently emphasizes effects of "othering" or *altération*: "the 'egoism' of an erotically reduced *ego* [is] that of an egoism altered from elsewhere, and opened by it" (25). Moreover, that *othering* is defined specifically as exposure to an elsewhere ("I am there where I find myself, there where a possible elsewhere affects me" [38]) understood as a function of the body, or more specifically of flesh ("where the elsewhere reaches me . . . [is] known by the name *flesh*" [38]), and, furthermore, implies a particular passivity that begins with touching—"auto-affection alone makes possible hetero-affection. . . . My hand only touches [another thing] precisely because it feels that it touches it; and it only feels that it touches it by feeling itself touch it, that is, by feeling itself at the same time as it feels what it feels" (114)—as well as a form of automatism of erotic excitation (138–43). On the other hand, the flesh that is aroused in such an erotic exposure is understood in the end not as exscription, as in Nancy, but via a surprisingly Christian sense of incarnation (221–22) and even of chastity: "Thus there opens before free eroticization an immense field of activity . . . where sexuality does not reach. From parent to child, from friend to friend, from man to God. Doubtless, we also recognize in free eroticism *chastity*, the erotic virtue par excellence" (183). I note in passing below other parallels between Marion and Nancy, or Barthes.

17 Georges Bataille, *Erotism: Death and Sensuality*, trans. Mary Dalwood (San Francisco: City Lights, 1986), 11.

18 Cf. *Corpus*: "An orgasm [*jouir*] is the diastole without systole at the *heart* of the dialectic" (39).

19 Jean-Luc Nancy, "The Intruder," in *Corpus*, esp. 165–68. See also Jacques Derrida, *On Touching—Jean-Luc Nancy*, trans. Christine Irizarry (Stanford, Calif.: Stanford University Press, 2005): "It is the thinking of a *technē* of bodies as thinking of the prosthetic supplement that will mark the greatest difference, it seems to me, between Nancy's discourse and other more or less contemporary discourses about the 'body proper' or 'flesh'" (97); and "At the heart of all the debates regarding the 'body proper' or the 'flesh'—ongoing debates as well as those that lie in wait for us—at the heart of the syncope . . . there is the originary intrusion, the ageless intrusion of technics, which is to say of transplantation or prosthesis. . . . Never again will the said 'technical' intrusion of the other, the ecotechnics of other bodies let themselves be reduced" (113).

20 For Thomas Wilson, *Arte of Rhetorique*, ed. Thomas J. Derrick (New

York: Garland, 1982), 1553; for my discussion, *Prosthesis,* 225ff.; for the artificial heart with continuous flow rather than the natural heart's pulse, see recent postings on http://www.texasheart.org/.

21   Roland Barthes, *A Lover's Discourse: Fragments,* trans. Richard Howard (New York: Hill and Wang, 1978), 40. Further references included in text are preceded by the mention *Lover's Discourse;* the translation is often modified.

See, by comparison, contrast, or extension, Deleuze on masoch(ism): "The woman-torturer sends a delayed wave of pain over the masochist, who makes use of it, obviously not as a source of pleasure, but as a flow to be followed in the constitution of an uninterrupted process of desire. What becomes essential is waiting or suspense as a plenitude." Gilles Deleuze, "Re-presentation of Masoch," in *Essays Clinical and Critical,* trans. Daniel W. Smith and Michael A. Greco (Minneapolis: University of Minnesota Press, 1997), 53–54; also Marion: "Elsewhere, as event, compels me to the posture of waiting [*attente*]. . . . Only my waiting lasts: it suspends the flux of time. . . . In the erotic reduction's time, there lasts only waiting, for which nothing happens" (*Erotic Phenomenon,* 33, translation modified; see also 35).

22   It would take a detailed analysis—and perhaps further modifications—to understand to what extent visual Internet communications such as Skype remain primarily aural or phonic: whether their photographic aspect is simply an extension of the letter, postcard, and mailed photograph, such as photographic sexting seems to suggest.

23   *On Touching,* 112–13.

24   See my discussion in *Dorsality,* 34–41.

25   Cf. Traditional *waiata* (Maori song): "Tangi a te ruru re kei te hokihoki mai e / E whakawherowhero re i te putahitanga / Naku nei ra koe i tuku kia haere / Te puritia iho, nui rawa te aroha" (The morepork's cry keeps coming to me / It is hooting out there where the paths meet / I was the one who allowed you to go / My great love did not detain you). The ruru plays an important part in Maori culture, from the eye motif in woodwork and in dancing (e.g., haka) to mythological associations (revered as an ancestral spirit; appearance near a dwelling the sign of death).

26   *Lover's Discourse,* 114, translation modified. Howard's translation unfortunately skips the all-important line regarding "inflection." Cf. *Fragments d'un discours amoureux,* in Roland Barthes, *Oeuvres complètes V: Livres, textes, entretiens 1977–1980* (Paris: Seuil, 2002): "La voix supporte, donne à lire et pour ainsi dire accomplit l'évanouissement de l'être aimé, car il appartient à la voix de mourir. Ce qui fait la voix, c'est ce qui, en elle, me déchire à force de devoir mourir, comme si elle était tout de suite et ne

pouvait être jamais rien d'autre qu'un souvenir? Cet être fantôme de la voix, c'est l'inflexion. L'inflexion, en quoi se définit toute voix, c'est ce qui est en train de se taire, c'est ce grain sonore qui se désagrège et s'évanouit. La voix de l'être aimé, je ne la connais jamais que morte, remémorée, rappelée à l'intérieur de ma tête, bien au-delà de l'oreille" (147).

27   See Deleuze's reading in *Proust and Signs*, trans. Richard Howard (Minneapolis: University of Minnesota Press, 2000), of this revelation of "a world of abomination" (125), in the context of his thesis that "essence, in love, is incarnated first in the laws of lying [*mensonge*], but second in the secrets of homosexuality" (80, translation modified; see also 9–11).

28   See Marion's brilliant likewise pages on how "erotic language not only speaks in order to say nothing, but most often says anything" (*Erotic Phenomenon*, 147), how "the erotic reduction and poetry are in league" (148), how erotic speech "becomes purely performative" (149), how lovers address one another "in turn with obscene words and with puerile words" (149), how erotic speech inevitably "borrow[s] from mystical theology" (149), corresponding an "affirmative path" (150, translation modified).

8. THE AUDIBLE LIFE OF THE IMAGE

1   I attempted briefly to outline this logic some years ago, with particular reference to *Every Man for Himself* and *First Name Carmen*, in "Representing Silence (in Godard)," in *Essays in Honour of Keith Val Sinclair*, ed. Bruce Merry, 180–92 (Townsville, Australia: James Cook University Press, 1991).

2   Cited in Jean-Luc Douin, *Jean-Luc Godard* (Paris: Rivages, 1994), 99–100, my translation and emphasis. The end of *Every Man for Himself* is inscribed within the same logic: "when Scarlett O'Hara is kissing Rhett Butler and you hear the hundred violins of the Boston Symphony Orchestra, or when the Russian Army crosses and we hear the hundred violins of the Chicago Orchestra . . . we want to *see* the music." Jean-Luc Godard, "A New Direction for the New Wave's Jean-Luc Godard," interview with Annette Insdorf, in *Jean-Luc Godard: Interviews*, ed. David Sterritt (Jackson: University of Mississippi Press), 89. See also, e.g., Jacques Aumont: "Godard is interested neither in the 'grand form' nor the musical phrase for itself, but in the musical *idea*." Aumont, "Lumière de la musique," *Cahiers du cinéma* 437 (1990): 46–48, my translation.

3   Adrian Martin, "Recital: Three Lyrical Interludes in Godard," in *For Ever Godard*, ed. Michael Temple, James S. Williams, and Michael Witt (London: Black Dog, 2007), 255.

4   Deleuze and Guattari, *What Is Philosophy?*, 2.

5   See also Deleuze and Guattari, *What Is Philosophy?*: "With the means
provided by its materials the aim of art is to wrest the percept from perceptions of objects and the states of a perceiving subject, to wrest the affect
from affections as the transition from one state to another: to extract a
bloc of sensations, a pure being of sensations" (167, translation modified; cf. *Qu'est-ce que la philosophie,* 158). Also, "the affect goes beyond
affections no less than the percept goes beyond perceptions. The affect
is not the passage from one lived state to another but man's nonhuman
becoming" (173).

6   Cf. Gilles Deleuze, *L'Image-temps: Cinéma 2* (Paris: Minuit, 1985), 111.

7   The translation has "moments" for "movements" ("au-delà meme des
mouvements du monde"); cf. ibid., 108.

8   Claire Colebrook, *Deleuze: A Guide for the Perplexed* (London: Continuum, 2006), 4.

9   Ibid., 5.

10   Claire Colebrook, *Deleuze and the Meaning of Life* (New York: Continuum, 2010), 20.

11   Ibid.

12   D. N. Rodowick, *Gilles Deleuze's Time Machine* (Durham, N.C.: Duke
University Press, 1997), 139.

13   Ibid., 201.

14   Ibid., 140.

15   See also *Cinema 2,* 234–35, and my discussion in "Representing Silence."

16   A flawless "spatio-temporal integrality," where on-screen time equals real
time, as in a theatrical scene. Christian Metz, *Film Language: A Semiot-
ics of the Cinema,* trans. Michael Taylor (New York: Oxford University
Press, 1974), 129.

17   Douglas Morrey, *Jean-Luc Godard* (Manchester: Manchester University
Press, 2005), 77.

18   Raoul Coutard comments on the practical problems of producing the famous traffic jam shot in Colin MacCabe's "Interview with Raoul Coutard"
on the DVD of *Week-end* (United States: New Yorker Films, 2005).

19   As a prime example, see *Vivre sa vie* [My life to live] (1962), Tableaus 1
and 7.

20   George Bataille, *Story of the Eye,* trans. Joachim Neugroschel (San Francisco: City Lights, 1987), 4–5, 10–11, 5–40. *A la cuisine les minettes* literally
means "(let's go) to the kitchen, babes," but *la cuisine* can also refer to
where things (sexual and otherwise) get cooked or heated up.

21   See Godard's explanation concerning having only two hands, hence being
able to use only two tracks of sound at any given time: "all my films since
Sauve qui peut have only two mixing tracks. You know, in the cinema,

when it comes to making up the final sound track, there are always many original tapes—the sound of a car arriving at the beach, for example, the voices of the actors getting out of it and saying 'I love you' or the opposite, the sound of the waves behind them, maybe a cock crowing in a nearby farm, and some music. That makes five tracks . . . and I have only two hands to manipulate them. . . . If I had only one arm, maybe I'd have only a single sound track. All this business of having the various sounds marching up in front of you like soldiers is called mixing and I can only control it with the two hands I have." Gideon Bachman, "The Carrots Are Cooked: A Conversation with Jean-Luc Godard," in Sterritt, *Jean-Luc Godard: Interviews,* 134.

22 In this respect, she recalls Brigitte Bardot in *Le Mépris (Contempt),* lying naked, playing with her hair in a similar way while speaking to her husband, and in both cases Godard has wrung from commercially popular actresses whom he didn't want performances of considerable, and understated, emotional depth.

23 Bataille, *Story of the Eye,* 40, translation modified.

24 Cf. Georges Bataille, "L'Amour d'un être mortel," in *Oeuvres complètes VIII* (Paris: Gallimard, 1976), 497, my translation. As Stuart and Michelle Kendall point out, Bataille reworks that text for *The Accursed Share,* trans. Robert Hurley (New York: Zone Books, 1991), 2:157–71; cf. Georges Bataille, *The Unfinished System of Nonknowledge,* trans. Stuart Kendall and Michelle Kendall (Minneapolis: University of Minnesota Press, 2001), 280n69. For further interesting discussion of Godard's use of the text, including in *Histoire(s) du cinéma,* see Douglas Morrey, "History of Resistance/Resistance of History: Godard's *Éloge de l'amour* (2001)," *Studies in French Cinema* 3, no. 2 (2003): 121–30.

25 See Georges Bataille, *Blue of Noon,* trans. Harry Matthews (London: Marion Boyars, 1988). Regarding Weil as model for Lazare, and their intellectual debates in general, see Alexander Irwin, *Saints of the Impossible: Bataille, Weil, and the Politics of the Sacred* (Minneapolis: University of Minnesota Press, 2002), 82–123. Irwin (82–86) discusses in detail the argument between Bataille and Weil over Malraux's *Human Condition* during their collaboration on the journal *La Critique sociale,* not long before Bataille began writing *Blue of Noon.*

26 Paul Celan, "Todesfuge" [Death fugue], from *Mohn und Gedächtnis* (1952), in *Poems of Paul Celan,* 30–33.

27 I am aware that a complex problematic arises concerning how to circumscribe the fragments from *In Praise of Love,* and later from *Notre musique,* that I am privileging here. I am exploiting a certain limit determined by the explicit sound track reference to, and quotation from, Bataille, while

also profiting from a contextual overflow that occurs, especially on the image track (but also, as we shall see, on the sound track). Whereas the audition scene is self-contained as a sequence, the framing instances are more complicated: the first occurs within a larger conversation that takes place in a car; the beginning and end of the second are difficult to determine. I note simply that (1) given Godard's use of montage, the terms *fragment* and *sequence* are themselves problematic, and (2) the idea of a musical outside to the image that I am developing here presumes a more radical fracturing of the filmic surface, such that the cohering unity of any larger or smaller grouping of images remains both arbitrary and unstable.

28   Laurent Jullier, "JLG/ECM," in Temple et al., *For Ever Godard,* 274; James S. Williams, "Music, Love, and the Cinematic Event," in ibid., 292.

29   Morrey, "History of Resistance/Resistance of History," 121, 123. Morrey does not clearly argue that it is in the unconventional love story that history comes to be found, preferring to see the latter's "recalcitrance" (124) analyzed in more conventional references—Nazism, the Shoah, and the Resistance; 1968; Kosovo. But he does establish a structural comparison in *In Praise of Love* between history's recalcitrance and the elusiveness of that love story, "resolutely opposed to the formation of the couple which tends to organize the Hollywood romance" (126), and suggests that idea "could perhaps be articulated around" the Bataille text (128), while declining to pursue the structural comparison into the State–love contrast developed by Bataille.

30   See Giorgio Agamben, *Remnants of Auschwitz: The Witness and the Archive,* trans. Daniel Heller-Roazen (New York: Zone Books, 2002), 41–86. Cf. Godard: "No one but me has said that at one point in the extermination camps the Germans had decided to declare a Jew to be a Muslim." Jean-Luc Godard and Youssef Ishaghpour, *Cinema: The Archeology of Film and the Memory of a Century,* trans. John Howe (Oxford: Berg, 2005), 105.

31   Godard and Ishaghpour, *Cinema,* 24.

9. MEDITATIONS FOR THE BIRDS

1   See, e.g., C. K. Catchpole and P. J. B. Slater, *Bird Song: Biological Themes and Variations* (Cambridge: Cambridge University Press, 1995), 9–11, 185.

2   "Our collection of birds are [sic] soft, cuddly and educational. Created in conjunction with the Audubon Society, each plays its own unique and realistic bird song, provided by the Cornell Lab of Ornithology." http://www.wildrepublic.com/en/audubon. The "gentle squeeze" quotation can be found on various retail websites.

3 In the interests of scientific rigor, I should note the following: (1) the experiments were carried out over a period of three months in early 2008; (2) the number of repetitions was determined by how many times daily my daughter, after having her diaper changed and examining first her snow baby (courtesy of Deeanna Rohr) and second her fish mobile (courtesy of Liana Theodoratou and Eduardo Cadava), cast her eyes longingly toward the birds (courtesy of Sharon Cameron); (3) since the very beginnings of my investigations, I felt compelled to give two squeezes to each of the four birds, producing two (or four) repetitions of the call, either because I sought to compound by a factor of two my daughter's pleasure or because there was some automatic impulse or desire at work in favor of a repetition of a repetition, a call and response effect, or for some other unconscious or unknown reason; (4) the "gentle squeeze" that the manufacturers cite was sometimes not enough to call forth the song, and a more complicated manipulation and variation of grip and pressure, at times bordering on violence, was required. However, never did I fail, in the end, to make each bird sing.

4 For example, in the debate between Foucault and Derrida, in Foucault, *History of Madness*, trans. Jonathan Murphy (London: Routledge, 2006), and Derrida, "Cogito and the History of Madness," in *Writing and Difference*, trans. Alan Bass (Chicago: University of Chicago Press, 1978).

5 Derrida comments, "The animal that I am not, the animal that in my very essence I am not, Descartes says, in short, presents itself as a human mind before naked wax" (*Animal*, 73).

6 Cf. Derrida, *Limited Inc.*, 2, 9; and Jacques Derrida, *Dissemination*, trans. Barbara Johnson (Chicago: University of Chicago Press, 1981).

7 R. E. Lemon, cited in Catchpole and Slater, *Bird Song*, 165. Preceding information is also from this source.

8 Aretas A. Saunders, *A Guide to Bird Songs* (New York: D. Appleton-Century, 1935), 241.

9 See Albert R. Brand, *Songs of Wild Birds* (New York: Thomas Nelson, 1934), 75, 79, and Albert R. Brand, *More Songs of Wild Birds* (New York: Thomas Nelson, 1936), 83.

10 Saunders, *A Guide to Bird Songs*, 104.

11 René Descartes, *Oeuvres et lettres*, ed. André Bridoux (Paris: Gallimard/Pléiade, 1953), 1004–5, my translation.

12 Lacan, "Function and Field of Speech and Language in Psychoanalysis," 84.

13 Ibid., 85.

14 Jacques Lacan, "Propos sur la causalité psychique," in *Écrits* (Paris: Editions du Seuil, 1966), 189–90.

15 See René Descartes, *The Correspondence,* in *The Philosophical Writings of Descartes, Vol. III,* trans. John Cottingham, Robert Stoothof, and Dugald Murdoch (Cambridge: Cambridge University Press, 1991), 41–43; cf. *Oeuvres I,* 286.

16 Descartes, *Correspondence,* 41.

17 See *Writings I,* 103, and Descartes, *Oeuvres et lettres,* 823, 865.

18 See http://www.cnrtl.fr/etymologie/poêle.

INDEX

264–67, 288–89n19, 289–90n24
"To Speculate—on 'Freud'"
 (Derrida), 56, 59, 73–77, 79,
 293n10
trace, 43–44, 52, 72, 84, 98–101,
 111, 122–25, 127, 133, 136–37,
 153, 178, 303n9; as *différance,*
 12–13, 90–91
Treatise on Man, The (Descartes),
 45, 257, 260, 274–76, 279–80

"Unconscious, The" (Freud), 57, 64

vitalism, 3, 7, 17–19, 25–27, 233

waiting: as condition of love,
 218–23, 228; and masochism,
 306n21
Week-end (Godard film), 230,
 235–43, 250, 253, 308n18
What is Philosophy? (Deleuze and
 Guattari), 289n19
wheel, and *autos* in Derrida: 90,
 101–3
Williams, William Carlos, 176–77

(continued from page ii)

David Wills is professor of French studies and comparative literature at Brown University. He has published extensively (as author, editor, coauthor, and coeditor) in French and comparative literature and in film theory, and he is a translator of and commentator on the work of Jacques Derrida.